Reproductive Behavior of Captive Nicobar Pigeons, *Caloenas nicobarica*

Revised Edition

Shauna S. Roberts
nonfiction: http://www.ShaunaSRoberts.com
fiction: http://www.ShaunaRoberts.com

NICOBAR PRESS
Riverside, California
http://www.NicobarPress.com

Nicobar Press
http://www.NicobarPress.com

First printing, June 2012

ISBN 978-1-938125-12-6

Front cover image is from Plate 10 in *Indian Pigeons and Doves* by E.C. Stuart Baker.
London: Witherby & Co., 1913.
Image is in the public domain and was obtained from Wikimedia:
http://commons.wikimedia.org/wiki/File:CaloenasNicobaricaBaker.jpg

Back cover image is from *The Royal Natural History,* Vol. 4, by Richard Lydekker.
London: Frederick Warne & Co., 1895.
Image is in the public domain and was obtained from Wikimedia:
http://commons.wikimedia.org/wiki/File:CaloenasLyd4.png

NORTHWESTERN UNIVERSITY

Reproductive Behavior of Captive Nicobar Pigeons, *Caloenas nicobarica*

A DISSERTATION

SUBMITTED TO THE GRADUATE SCHOOL
IN PARTIAL FULFILLMENT OF THE REQUIREMENTS

for the degree

DOCTOR OF PHILOSOPHY

Field of Anthropology

By

SHAUNA S. ROBERTS

EVANSTON, ILLINOIS

June 1984

ABSTRACT

Reproductive Behaviors of Captive Nicobar Pigeons, *Caloenas nicobarica*

Shauna S. Roberts

The Nicobar Pigeon (*Caloenas nicobarica*) is an endangered species native to small Indo-Australasian islands. The Lincoln Park Zoological Gardens in Chicago maintain a colony of Nicobar pigeons in a free-flight exhibit. This dissertation is based primarily on research conducted on this colony from 1980 to 1982, supplemented by information from preceding and subsequent years.

I observed the Nicobars during their breeding cycle. The courtship stage begins when the males, synchronously coming into breeding condition, choose nest sites and perform displays there. Females are attracted by the displays and join the males at the nest sites, where they may be either accepted or threatened away. When a female is accepted, the birds preen each other and may copulate.

The birds tend to pair with others hatched in the same year. Familiarity with the potential mate counts in its favor. Due to certain demographic characteristics, mates of the same age on average share fewer genes than random pairs, and produce less inbred young.

After courtship the pair builds a nest. Forked sticks longer than 12 inches are the preferred material. Shorter sticks and vegetation are concentrated in the top of the nest. Wide variations in stick preferences are evident.

Although the Nicobars defend their nests throughout nesting, threats are more likely to be given while building. Other species are usually tolerated; most threats are directed towards Nicobars. This may function to protect sticks from being stolen.

Both the male and the female incubate the egg and rear the chick. Behavioral development does not seem related to physical size.

Very little is known about the Nicobar pigeon either in captivity or in its natural habitat. Its further study and conservation are important. A colonial breeder, the Nicobar is in the minority among pigeons and can provide enlightening comparisons with the others. The unusual structure of its stomach is unique and merits further examination.

Preface

I first must thank Dr. Lester Fisher and the Lincoln Park Zoological Gardens for their cooperation in allowing me to do my research at the Zoo. Kevin Bell not only tolerated my presence at the Zoo off and on for over three years, but also assisted me in many ways. Kathleen Brown told me everything she knew about Nicobar Pigeons and kept me informed about their activities during my absences. She also allowed me to use Plate 7-1, provided some of the data, helped collect sticks, and aided in the compilation of Appendix II. Larry Kodani took photographs of the Nicobars, including Plates 1-1, 1-3, and 2-3 in this dissertation.

Partial support for this dissertation came from my aunt, the late Janet Louise Roberts, and from the Doctor Scholl Foundation and the Lincoln Park Zoological Society. Their generosity has been greatly appreciated. My advisor, Donald Stone Sade, always made sure that I had some form of support throughout graduate school, and provided me with a job for two years.

I want to thank all those who read the earlier drafts and papers that eventually were put together to form this dissertation. This includes the members of my committee, Donald Stone Sade, James Cheverud, Malcolm Dow, and Kevin Bell, as well as David Malueg and Kathleen Brown.

Laurie Rush listened, and gave helpful hints about writing the dissertation. I also thank Loni Korpel, who translated not only the Dutch in Bartels (1964) and Hoogerwerf (1967), but the Indonesian terms as well. Forrest Nelson helped me with covariance analysis.

Charles Moore typed this dissertation, and also drew Figure 2-2, for which I am very grateful.

An earlier version of Chapter 4 was presented at the 1983 Midwest Regional Animal Behavior meeting, and an earlier version of Chapter 6 at the 1982 Midwest Regional Animal Behavior meeting.

The biographies in Appendix II were inspired by and modeled after Donald Stone Sade's biographies of the rhesus monkeys of Cayo Santiago (Sade *et al.* in prep.).

My husband, David Malueg, provided a model of discipline and hard work to follow. He also provided love and encouragement and some statistical advice.

Table of Contents

updated 13 June 2012

List of Tables

List of Illustrations

Chapter One

The Nicobar Pigeon: A Literature Review

> Words fail to describe the beauty of this glorious pigeon. The lean, game-looking, iron-grey head and neck, the streaming ruff of long glittering hackles, the heavy body and strong broad wings of lustrous metallic green, changing in every light and set off by the square snow-white tail, all combine to make it one of the loveliest of known birds.
>
> Never shall I forget my first _Caloenas_!
>
> (Butler 1899:689)

Although few authors wax as poetic as in the above paragraph, a major part of the literature on the Nicobar pigeon consists only of physical description or description of range. In this chapter, I will summarize the literature with a greater emphasis on behavior.

Linné (1758:164) named the Nicobar pigeon _Columba nicobarica_. He described it briefly as having a white tail, a black body, and a shining green back, and listed its habitat as the Nicobar Island (sic). Salvadori (1893:615) cites the earliest use of the generic name _Caloenas_ as 1840, and gives 16 other Latin names by which the pigeon was also known.

Various authors give the length of the adult Nicobar pigeon as 16 inches (Ali & Ripley 1981:161), 14 inches (Salvadori 1893:617), 15-1/4 to 16-1/2 inches (Hume 1874:271), 15 to 16-1/2 inches (Baker 1928:214) and 12-1/2 to 13 inches (Rand & Gilliard 1967). The adult weight is 1.3 pounds (Heinroth 1902:408), 1.25 to 1.75 pounds (Hume 1874:271) or 1 to 1.3 pounds (Ali & Ripley 1981:162).

Plate 1-1

Adult Nicobar Pigeon

Physically, the Nicobar (Plate 1-1) has been described as:

> [a] dark slaty grey pigeon with lustrous metallic blue-green and copper-bronze upperparts; unmistakable by the glistening mane-like neck hackles and sharply contrasting white tail-coverts and tail. (Ali & Ripley 1981:161)
>
> Kopf, Hals, ganze Unterseite und Schwingen schwarzgrün, alle Federn der Unterseite cyanblau gesäumt; die untersten längsten Halsfedern, die am Kragen, Rücken, Bürzel und der Flügeldecken grasgrün metallisch, auf dem Kragen goldglänzend; Unterflügel schwarzblau; Schwanz weis; Schnabel und Iris schwartz; Beine dunkelroth, Zehen unterseits und Nägel gelb. (Reichenbach 1874:48)
>
> Heavy bill and rather vulture-like stance and carriage. Feathers on head very short, those on front of neck and upperpart of hind neck somewhat elongated and hairy in texture. Feathers elsewhere on hind neck elongated to form long shining hackles .... Head and the hairy upper neck feathers dark blackish grey with a silvery purple bloom .... Primaries and secondaries dark blue, glossed more or less with green and purple, inner webs blackish. Tail white. Upperparts shining dark green or dark green shot with coppery red. Underparts dark, iridescent green or green and blue .... Irides brown, sometimes whitish. Bill and cere blackish or dark grey. Feet purplish or dark purplish red with yellow or pale brown claws. (Goodwin 1970:262-263)

Other authors (Rand & Gilliard 1967; Baker 1913:114) go into similar detail. Hume (1874:271-272) devotes nearly two full pages to its description but concedes that:

> It is as difficult to convey in words any adequate description of the ever-changing tints of this bird as it would be to paint the sparkle of a diamond. (Page 272)

Plate 1-2

Immature Nicobar Pigeon

Hume (1874:271) says the sexes are very similar although females may have shorter hackles. Ali and Ripley (1981:161) and Baker (1913:115; 1928:214) claim females have greyer heads, necks, and breasts, have lighter blue upperparts, and shorter hackles. Rand and Gilliard (1967) believe the females are more brownish and have longer tails with dark tipping. Goodwin (1970:263) says the female is smaller, with correspondingly smaller cere and hackles, and has more copper color.

The juvenile Nicobar pigeon (Plate 1-2) is described by Goodwin (1970:263) as like a female but duller and hackle-less, with a blackish green tail and blackish feet. Reichenbach (1847:48) agrees that in the juvenile the hackles are missing and the tail is blackish green. Baker (1913:115) says the young male has no hackles, the tail matches the upperparts, and the eyes are hazel-brown.

The above descriptions are presumably of _Caloenas nicobarica nicobarica_. The other subspecies, _C. n. pelewensis_, is supposedly very similar, but is slightly smaller and bluer with shorter neck hackles that sometimes bifurcate (Goodwin 1970:263).

The Nicobar is found "on small, wooded coral islands of the Indo-Australian archipelago and the southwestern part of the Pacific Ocean" (Bartels 1964:97). More specifically, _C. n. pelewensis_ has been reported as rare or uncommon on six of the larger Palau Islands (Pratt, _et al_. 1980:119) and is considered endangered (King 1981). _C. n. nicobarica_ occurs on both of the Surin Islands, but is rare (Brockelman & Nadee 1977:224). It occurs and is more common on several of the Nicobar Islands (Ali & Ripley 1981:161) where the heaviest concentration of breeding birds are found on Batty Malve (Hume 1874:271), also known as Battye Malve, Batti Malv, and Batti Malve. Allen (1951:154) found it on one of the Sembilan Islands. Rand and Gilliard (1967) additionally list as part of the range the Mergui Archipelago, the Sunda Islands, the Philippine Islands, the Molucca Islands, the Bismarck Archipelago, the Solomon Islands, the Pecan Islands, the northern Papuan Islands, the islands of Geelvink Bay, the Trobriand Islands, Normanby Island, the Woodlark Islands, and the Louisiade Archipelago. The Cocos and Andaman Islands are also inhabited by Nicobar pigeons (Baker 1928:213). That the Nicobar pigeon is found "only on small wooded islets off the larger land masses" (Goodwin 1970:263) is generally agreed upon, except that Wallace (1865:370) found it upon mainland New Guinea, as did Hoogerwerf (1967:249). These islands are mostly uninhabited and free from predators (Wallace 1865:370, 395), and are all tropical, lying between 20° N and 15° S latitude.

The Nicobar pigeon's stomach is unique. Flower (1860) described the two very hard circular bodies in the gizzard, not found in any other species. These powerful plates can grind up the hard pits and stones of fruit (Heinroth 1902:408). Between the two hard plates is frequently a stone, first noted by Dawson (Hume 1874:275). The Malays use this stone in their medicine (Wolfe 1951:204). Davison (Hume 1874:275) described the stone as a pea-sized pebble of white quartz. Bartels (1964:93) found one bird out of the ten examined had no stone, while others had siliceous, "often beautifully colored, hard semi-precious" stones that aided in grinding up seeds.

The 4000 mile range of the Nicobar pigeon is very large for a land bird (Wallace 1865:370; Wallace 1869:264). Wallace (1869:264-265) wondered at the huge pectoral muscles and large wings (Plate 1-3) and concluded that the function was to allow occasional migration. He knew of a case in which a Nicobar pigeon had flown a hundred miles. Yet he believed:

> The majority of individuals, therefore, can never make full use of their enormously powerful wings, till the exceptional case occurs of an individual being blown out to sea, or driven to emigrate by the incursion of some carnivorous animal, or the pressure of scarcity of food. (Wallace 1869:265)

Today much more frequent wanderings are believed to occur, although no one knows why they occur or how far the birds travel (Hoogerwerf 1967). Possibly Nicobars migrate from the largest islands to the smaller islands to breed each year (Rand & Gilliard 1967; Mayr 1945:66). Pratt (Pratt _et al_. 1980:122) watched a tiny group of Nicobars fly each afternoon from one island to another. Major long distance migrations are unlikely to occur this often. For example, no Nicobars were seen on the Palau Islands for 65 years; this record was broken by sightings of seven birds in 1945 on three different islands (King 1981). Nicobars have inhabited Palau since. These seven may have been migrants, or they may have been remnants of a Nicobar population so decimated that they eluded human detection. In either case, for decades no Nicobars successfully colonized Palau.

As might be expected from the preceding section on the stomach, the ground feeding Nicobars eat fallen, fleshy fruits, seeds (Ali & Ripley 1981:162; Bartels 1964:93), and possibly occasional invertebrates (Goodwin 1970:263). On the ground they walk about, with no apparent goal or pattern, every now and then digging into the dirt, or pushing aside leaves to look for food (Hume 1874:274; Ali & Ripley 1981:162; Baker 1928:214).

The Nicobars form groups of a variety of sizes. Davison (Hume 1874:273-274) once saw a group of at least 30 Nicobars; at other times he saw lone birds, pairs of birds, and sometimes groups of six to twelve. Butler (1899:689) saw both single birds and groups of three to twelve. In the 1800's, group size apparently increased dramatically during breeding, when up to thousands of Nicobars crowded onto tiny islands to breed (Hume 1890:365).

Nicobar reproduction occupies the most space in the behavioral literature, most of the information being drawn, however, from a very few sources. Davison (Hume 1874:274-275) believed himself to be the first to report accurately on Nicobar breeding. He discovered that the Nicobars congregated in large numbers on Batty Malve and built their nests in thick bushy trees, up to 13 per tree. The nests were typical pigeon nests -- a twig platform with no lining -- and each contained only one egg or chick. He found one nest only 10 feet off the ground, but the rest were 20 to 30 feet high. Hume (1890:365) claimed Batty Malve to be the only island of the Nicobar chain on which the Nicobars bred. Later a few Nicobars were also found to breed on South Sentinal Island (Baker 1913:115). Baker (1913:116) described the eggs as long, regular ovals, with an average length of 1.95 inches and an average width of 1.34. He also reported that at the Calcutta Zoo, the Nicobars had single chicks, fed them regurgitated food, and split incubation and care duties between the two parents. According to Baker (1928:214) the Nicobars breed in March in the wild and build their nests 10 to 30 feet off the ground. Ali & Ripley (1981:162) define the breeding season as "January to March-April" and also say that the nests are located 10 to 26 feet off the ground.

Knowledge about other Nicobar behavior is fragmentary. Non-feeding time is spent sitting on high horizontal branches (Butler 1899:689). Their vocalization has been described as a "croak, croak" (Hume 1890:365), a grunt like a pig (Whitman 1919:142), and "ein kurzes, teifes aber leises Rücksen" (Heinroth 1902:409). At the Calcutta Zoo, the Nicobars followed the visitors, waiting for treats, and got along well with their cagemates, both conspecifics and other birds. The males bow as other pigeons do (Baker 1913:118). In the wild, two birds meeting while foraging pecked once or twice at each other (Hume 1874:274). Davison (Hume 1874:274) observed:

> Occasionally a young one would rush up with outspread wings to one of its neighbours, and then stand with open mouth flapping its wings till it was either beaten off, or the other beat a retreat, but I did not see any of the young fed by their parents.

Two Nicobars observed on Palau circled each other soundlessly, their wings flapping and their hackles erect (Pratt et al 1980:122). Goodwin (1970:263) described two displays he observed performed by a captive male Nicobar. In the first, the Nicobar paraded back and forth, its head slightly lowered, the wings partially opened and lowered. In the other display the Nicobar bowed his head quickly and so low that the bill seemed nearly horizontal but upside down.

Nicolai (1962:127, 129) found in his captive colony of Nicobar pigeons that they did not dust bathe (he was unable to find out about rain and sun bathing) and that they were pronouncedly crepuscular and nocturnal. Whitman (1919:142) maintained at least one Nicobar pigeon in captivity, and described its threat display:

> When a strange pigeon ... approaches on the floor of the pen, the Nicobar pigeon (Cal. nicobarica) raises its neck and back feathers, reaches forward with its head with beak open, threatening to bite (not to peck). It opens the beak and "grunts" as it "throws" its head towards the intruder with beak open. The head keeps moving as if the bird was trying to reach the opponent, but the purpose served is to frighten off the other bird. If the intruder comes quite near there probably will be no actual biting, but there may be a retreat or a defense by striking with the wing. The grunt sounds much like that of a pig ....

The evolutionary history of the Nicobar pigeon is unknown. One third of extant pigeons live in the Malay Archipelago region, an area lacking in frugivorous competitors such as squirrels, monkeys, and tanagers (Wallace 1865:365-366). Both Wallace and Mayr (1946:18) believed columbids evolved in the Old World, possibly the Australian area. Wallace (1865:370, 395) thought the Nicobar pigeon originated on New Guinea and spread out from there to other predator-free islands. The Nicobar appears no more closely related to any particular genus of pigeons than to others (Goodwin 1970:262), and is considered one of the most "aberrant" pigeons (Mayr 1946:18). Thus, of its evolution it can only be said that it split off early from the other pigeons and remained in the pigeon homeland of the Pacific islands.

Plate 1-3

Nicobar Pigeon in Flight

Chapter Two

Study Site, Methods, and Goals

I studied the Nicobar pigeons in the Bird House at the Lincoln Park Zoological Gardens, Chicago, Illinois. The Bird House contains a free-flight exhibit (Plate 2-1) housing some 75 individuals of about 30 species, in addition to the Nicobars (Bell 1981:217). The free-flight exhibit is 102 feet long by 32 feet wide and averages 30 feet high, and contains many plants and trees (Figure 2-2). It is lit by natural light, and an artificial stream runs through it. Showerheads in the ceiling provide "rain" that can be turned on or off at the keepers' discretion. As the free-flight area is barless, the Nicobars also have free access to the visitor walk areas (roughly 59,000 cubic feet), to keeper walkways (roughly 15,000 cubic feet), and to a smaller free-flight area in the rear of the building (roughly 42,000 cubic feet). Although on rare occasions the temperature may rise into the low 90's or fall overnight into the low 50's, it usually is maintained around 70° to 75° F (Kevin Bell, personal communication).

The Zoo provisions the Nicobars with Purina Duck Growena, Purina High-Protein Monkey Chow 5045, and bowls of seeds and nuts (Bell 1981). They also eat grapes and peanuts. During this study, food was put out about 1000 hours and removed from the exhibit about 1630 hours.

Plate 2-1

Lincoln Park Zoo Bird House Free-Flight Area

FIGURE 2-2

Map of Lincoln Park Zoo

Bird House Free-Flight Area

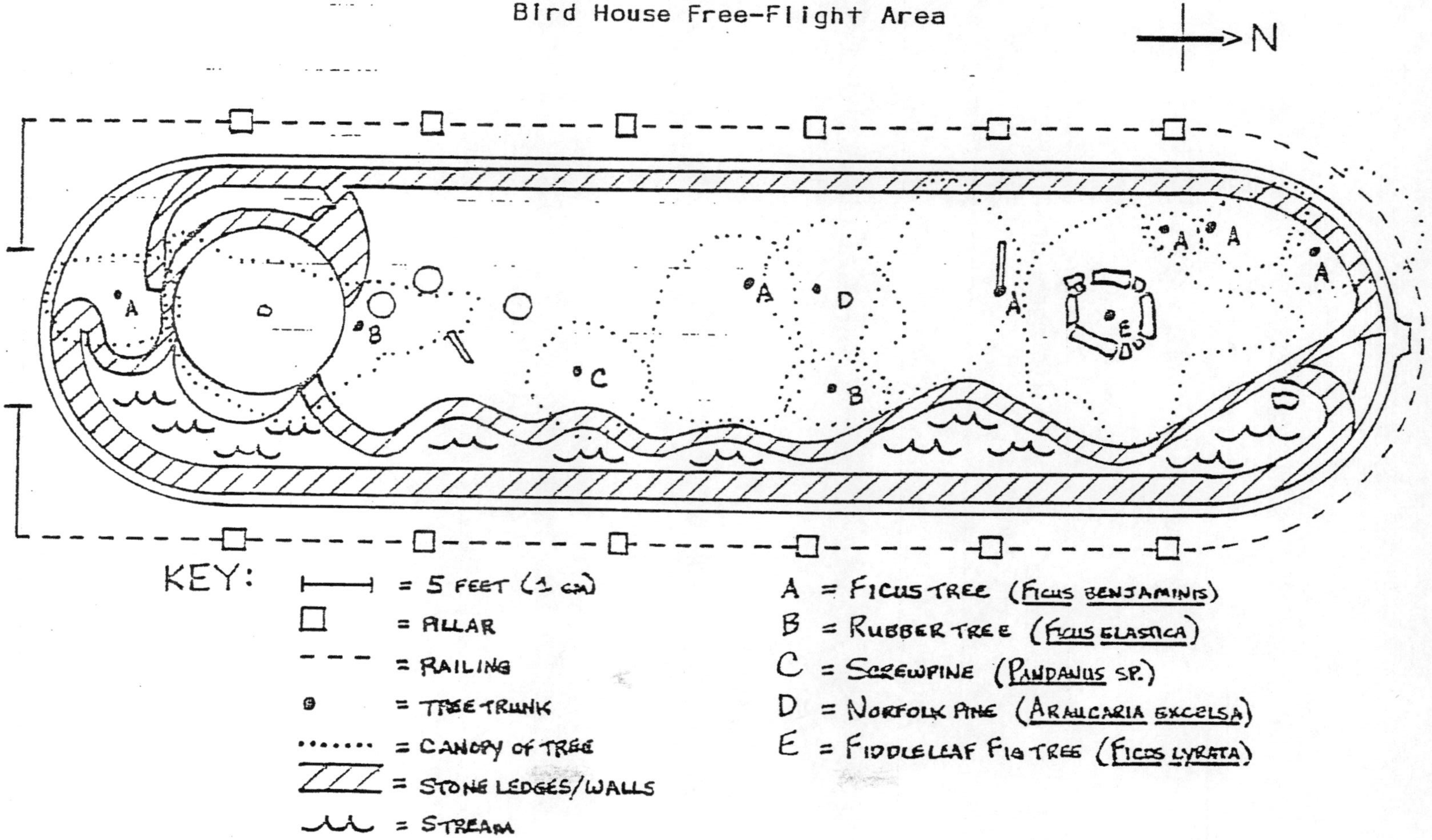

Table 2-1

Demographic Composition of the Colony

Date	males age 1 or older	females age 1 or older	immatures under 1	unknown sex 1 or older
12/31/76	4	3	2	
12/31/77	4	5	6	
12/31/78	6	7	9	1
12/31/79	8	7		1
12/31/80	6	5	8	
12/31/81	7	9	2	
12/31/82	5	6	7	1

Plate 2-3

Nicobar Pigeon Group

The history of the colony has been recorded since 1976, when Kevin Bell became curator. At that time, three males and three females comprised the colony. Three of these birds probably arrived at the Zoo in January 1973, one arrived in May 1975, and the history of the other two is unknown. Two more adults (one male, one female) were added in September 1976, an adult male spent six months at Lincoln Park Zoo in 1979, four Nicobars (including one adult male and one adult female) came in May 1980, and an adult female was added in September 1983. The 59 Nicobars in Appendix II include these 14 Nicobars and their descendants. Of the 59, 20 have been sent on breeding loans to other zoos and 14 birds died post-fledging between 1976 and 3 January 1984. The demographic composition of the colony is detailed in Table 2-1. For more information on additions to and removals from the colony, please see Appendix II.

The Lincoln Park Zoo has been very successful in breeding the Nicobars. Between April 1976 and December 1983, 97 eggs were laid, of which 44 resulted in fledglings. Cumulative egg production ranged between 0 and 28 eggs per bird.

This dissertation is based upon data collected between 9 June and 28 August 1980, 4 April and 21 May 1981, and 16 February and 10 August 1982. In 1980, I spent 17 days at the Zoo, making 84 hours of observations of the Nicobars. In 1981, 6 days resulted in 20 hours of observation. In 1982, 102 days were spent at the Zoo, with 521 hours spent observing Nicobars. In addition, many hours were spent taking photographs, collecting sticks, taking censuses, making a map, and examining the Zoo's records.

I took notes by ad libitum sampling (Altmann 1974:235) and focal animal sampling (Altmann 1974:242). Focal animal sampling was used to study group composition, individual activities, and fledglings; *ad libitum* sampling, usually modified to concentrate on one nest or pair rather than the whole group, was used for all other observations. In 1980 and 1981, I recorded observations in longhand in notebooks. In 1982, I used both notebooks and cassette tape recordings of spoken notes. Notes were primarily on reproductive behaviors, especially those concerned with constructing a nest. I used 7 X 26 binoculars for some observations. Since the Nicobars wear colored plastic and/or aluminum leg bands, I usually was able to identify the individuals I observed.

The Zoo proved a favorable environment for observations. All the birds were marked and their histories known. The free-flight area was small enough that each day the entire area could be examined, and any dead birds or new nests or hatchings could be discovered quickly. Autopsies were occasionally possible through the Zoo's veterinary facility. The Nicobars were quite used to having people around, overcoming the problem of acute shyness encountered by other observers (Nicolai 1962:129; Butler 1899:689; Pratt *et al*. 1980:122). Also, the Zoo's normal record keeping on the Nicobars (which recorded pairs, locations of nests, weights of fledglings, dates of egg-laying, hatching, and fledging, and miscellaneous other information; see for examples the pre-1980 data in Appendix II) enabled me to be absent without missing everything that happened.

There were a few disadvantages to the Lincoln Park Zoo as a study site. One minor disadvantage was that the Zoo had a commitment to breeding as many of the rare Nicobar pigeons as possible. I needed to choose projects that would not interfere with this goal. A larger problem was the massive crowds, especially on weekends and sunny days. I usually observed from the public area because visibility and ease of movement were better than within the free-flight area. Answering the expected questions such as "where are the elephants?" and "where are the restrooms?" was not distracting. But dealing with people who shoved their way in front of me and blocked my view, children who chased the pigeons or climbed into their enclosure, and individuals who made sexual advances sometimes did distract me and caused me to miss important behaviors. A final problem was the almost unaided natural lighting; when it stormed outside, observations became impossible inside. These problems are exceedingly minor compared to those faced by observers under more natural circumstances.

It seems apropos to now compare the Lincoln Park Zoo Nicobars with the birds described in Chapter 1. In physical appearance the birds resemble the descriptions given earlier; I shall not try to rival those authors' flowery phrases. Our males and females are for the most part not distinguishable by eye without reference to their legbands. Males are heavier at fledging than females. A two-tailed t-test of the difference between the average weights at fledging of males and females adjusted for age through covariance analysis (Snedecor & Cochran 1968:419-435) showed them to be significantly different ($t_{.05,19} = 2.093$), with males being heavier. As adults the neck hackles of males seem to be longer on average than those of females (Brown 1983). Adult male weights range between 1.18 and 1.87 pounds, female weights between 1.21 and 1.63 pounds. The Nicobars thus are heavier in captivity than in the wild. Our young birds, like those described in Chapter 1, lack hackles, have black tails, and are less irridescent (Plate 1-2).

Group size of course has as the upper bound the size of the colony. Foraging Nicobars (looking much like those described earlier) tend to walk alone or with one other bird, although many might forage at the same time. During two weeks before a breeding season started (16 to 26 February 1982) I made 34-1/2 hours of observation on group compositions and individual activities. Groups (defined as a local clustering of pigeons surrounded by a large area with few or no pigeons) mainly occurred under the southern rubber tree and around the ficus tree root south of the fiddleleaf fig (see Figure 2-2). Pigeons within groups (Plate 2-3) rarely interacted, but rather sat quietly, stood quietly, looked around, groomed themselves, or foraged (see Appendix I). The largest group I saw during this time had 14 individuals, consistent with group sizes reported earlier in Chapter 1.

In the Zoo, reproductive behaviors resembled what others had observed (and will be expanded upon in later chapters) except for two points:

1. Nests were lower at the Zoo. The lowest were roughly 5 feet high, the highest no more than 25 to 30 feet. Ceiling height limited tree height.

FIGURE 2-4

WEIGHTS OF MALE AND FEMALE FLEDGLINGS

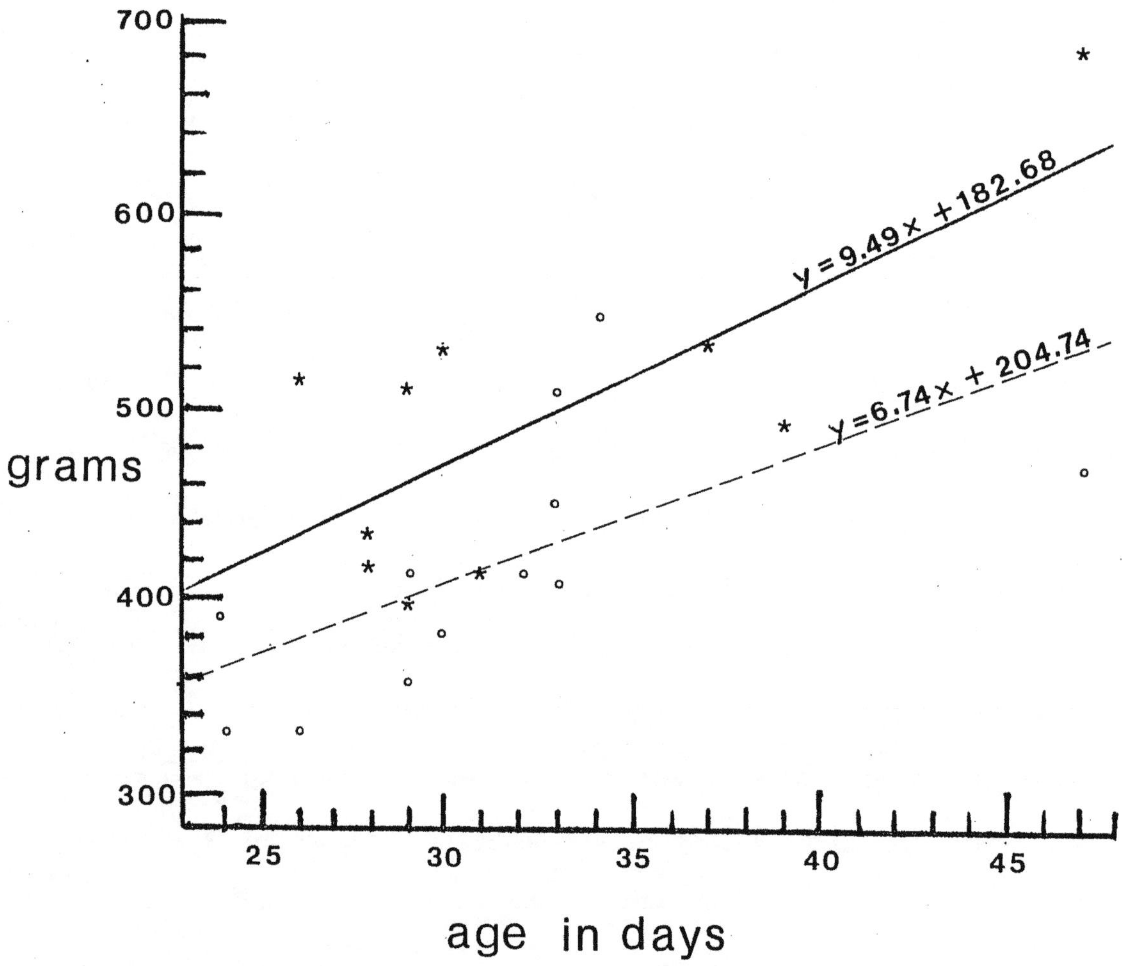

* = male ⸻ = male regression line

∘ = female - - - = female regression line

2. The peak egg laying months are April and May, with breeding occuring all year round.

I heard only one Nicobar vocalization -- a very soft, almost mechanical humming (mmrrrrr mmrrrrrr mmrrrrrr).

As described in Chapter 1, the Lincoln Park Zoo Nicobars also got along peacefully together and with their cagemates (Bell 1981:217).

Our Nicobars did not dust bathe. However, on many occasions I saw them rain bathing, and a few times sun bathing. Also in contrast to Nicolai (1962), ours are active diurnally. Whether they are active at other times as well is not known.

I observed all the behaviors and displays described on pages 9 to 10. Most will be described in Appendix I, and discussed in later chapters.

Thus, based on the limited reports of Nicobar appearance and deportment, the Lincoln Park Zoo Nicobars are much like their wild relatives. This lends validity to the following results, making it more likely that the behaviors discussed are more than distortions induced by captivity.

The Nicobar pigeon is on the path to extinction. They taste good (Baker 1913:118) and their formerly safe islands are now accessible to humans with motorboats (King 1981). Their future prospects seem dim. Because birds adapted to small island life have great difficulty colonizing larger islands or mainlands (Salomonsen 1975:590-591), any effort to establish colonies in wildlife refuges or parks will probably fail. Some scientists believe the coming century will see mass extinctions rivaling those of the dinosaurs; for example, 66% of New World tropical bird species may disappear (Lewin 1983). The Nicobar is at an extreme disadvantage compared to many other species. It is large, has a low reproductive rate, and has specialized ecological requirements -- all characteristics of bird species most vulnerable to extinction (Temple 1981:152).

One hope for their preservation is breeding programs in zoos. Although Nicobars were first bred in captivity over a century ago (Hopkinson 1926:117), current breeding programs are far from producing abundant numbers. Therefore the major and critical goal of this study is to provide as much information as possible on reproductive behaviors, with the hope of aiding efforts to breed the Nicobars. Not every reproductive behavior is discussed here. Some behaviors could only be watched at the expense of others (for example, my focus on the individuals on the nest prevented me from seeing their partners' searches for sticks), and others occurred too infrequently or too synchronously for me to get a good sample. The questions covered include:

1. What is Nicobar courtship like?
2. How do Nicobars choose mates?
3. How do the Nicobars construct their nests?
4. What types of sticks do Nicobars prefer?
5. What type of territoriality do the Nicobars have?
6. How do the Nicobars raise their chicks?

Two appendices provide more information. Appendix I is an ethogram of the Nicobar pigeon. It lists Nicobar behaviors, especially reproductive behaviors, describes each, and discusses their variation, if known. Appendix II contains biographical information -- hatchdate, parents, offspring, mates, medical information, weights, and dispositions -- on each Nicobar who spent any time as a post-fledgling in the colony. The records on which Appendix II is based are not always complete and are sometimes contradictory. I have made the best judgements I could and believe the information to be as accurate as possible, given the nature of the records on which it is based.

Summary

At the Lincoln Park Zoo Bird House lives a small colony of Nicobar pigeons in a free-flight exhibit featuring trees, natural light, and an artificial stream. Records of these pigeons have been kept since 1976, and may be viewed in abbreviated and collated form in Appendix II. I collected data in 1980, 1981, and 1982 on nestbuilding and other reproductive behaviors. The Lincoln Park Zoo Bird House turned out to be an excellent place for making observations. Nicobar behaviors seen in the wild (described in Chapter 1) were also seen at the Zoo. A main purpose of this study is to aid captive breeding programs to prevent this unique pigeon from becoming extinct. Appendix I contains a catalog of behaviors.

Chapter Three

Courtship

Courtship among the Nicobar pigeons resembles that of other pigeons but is more dramatic. The Nicobars come into breeding condition synchronously. Suddenly the free-flight area seems full of males treading in trees; then the birds are paired up; and then the males are all searching the ground for sticks to take back to females guarding nest sites. This is quite frustrating for the observer, who one week is forced to choose among several courting pairs, and the next week be hard put to find any.

During my 1982 observations at the Zoo, three cycles took place. I observed courtship from the beginning of each cycle until at least one pair began constructing their nest. After nestbuilding began, I still saw and made some notes on courtship.

Most observations were made between 1 and 10 March, 15 and 26 April, and 14 and 19 June. Several times a day I circled the free-flight area, checking each tree and noting any displays and pairs of Nicobars. In between I watched pairs. I gave preference to watching new pairs, inexperienced birds, and unusual events (such as homosexual behavior).

During these three cycles the population numbered from 17 to 19. Two adult males and one adult female died from an unknown disease that has since killed many Nicobars. All Nicobars except two surviving 1983 chicks participated in courtship.

Not all the participating Nicobars cycled together all three times. Pairs already with eggs or chicks did not court or lay new eggs (with one exception -- see Chapter 7). This absence of clutch overlap is surprising, given the similarity of Nicobar and feral pigeon parental care (Burley 1980:232-233). However, it may be due to the more frequent mate changes among the Nicobars (Burley 1980:230) or to their synchronous cycles.

I define the courtship period as the time between a male's first displays and the beginning of nestbuilding with his mate. This androcentric definition is used because male courtship behaviors are much more dramatic and obvious than those of females.

Several behaviors happen during courtship. They include:

1. Treading
2. Bowing
3. Head Throwing
4. Nodding
5. White Tail Fanning
6. Ritual Looking
7. Following
8. Begging
9. Lowered-Wing Walk
12. Mounting
13. Allopreening

These are described in Appendix I.

Courtship consists of two phases. In the first phase a mate is selected. The second phase is an "engagement period" during which the bond is cemented, but partners can still be changed before work on the nest begins.

In three 1982 breeding cycles, the initiation of the mate selection phase was obvious: each time, one morning I found the Nicobars up in the trees. (Usually they spend most of their time on the ground.) In the past, the start of courtship has also been signaled by a shift to the trees (Dennis Pate, personal communication). During the mate selection phase, males stand in the trees, in crotches or on wire platforms, bowing. In between sequences of bows, the males tread, throw their heads, and/or nod. The bows usually are not oriented towards anyone; there may be no other Nicobars in sight. Some males alternate their bows between two or more potential sites. Sometimes too they may flap their tails while at the platforms.

As a male displays, sometimes a female may approach. She stands or sits near him, not looking at him, sometimes preening herself. The male may bow towards her or in another direction. If the female finally approaches within reach, the male stops bowing. Sometimes they preen each other. Sometimes he threatens her or tries to bite or peck her. If attacked, the female usually does not move far, but returns when the attack stops, presumably demonstrating the strength of her interest (see Trivers 1972:172). Allopreening may replace attacking in her future approches. But sometimes the male always attacks her or avoids her, no matter how many times she approaches.

Females may visit several displaying males, or concentrate their attention on just one. Their approaches are usually characterized by hesitancy. The female may alternate moving closer to the male with backing off, or she may walk sideways towards him. I once saw the same movements used by a Nicobar, presumably afraid, approaching a peanut under a chair on which I was sitting. Presumably then the female approaching the male at his site is apprehensive.

Here are paraphrases from my field notes with examples of female approaches:

> 16JUN82 0912 Male 86 nods at site, female 66 flies over and lands a couple of feet above him. Then she walks around. 0915 66 is now standing right next to 86. He thrusts his open beak at her but she stays there. Then she preens 86's back.

> 15JUN82 0908 Male 86 bows, then spreads his white tail and fans it up and down 12 times. Female 66 moves into the rubber tree, far from 86. 86 bows. 0910 Female 35 walks through the nest basket and approaches 86. She hesitates for a long time (stepping around, moving closer and farther away), then she hops up next to 86. 86 now throws his head up. 35 turns around a few times. 0911 They preen each other.

> 4MAR82 1135 Male 48 flew over to and landed on the East Run handrail. Female 35 walks towards him. He pulls out one of her feathers, and she turns and walks away. Then she comes towards him again, and begs as he threatens her. He does not feed her.

> 20APR82 0949 Male 68 stands a foot from the 22/23 nest, then he bows. Female 41 approaches and sits next to him. She approaches him again. They sit facing the same direction, preening each other.

Nicobars also interact with prospective mates away from the nest platforms. One Nicobar may follow another on the ground or on the handrails. Possibly this is the male's chance to take a more active role in choosing a mate by following a female who has not visited his site. A female may follow the same male on the ground that she has approached at his site.

Sometimes more than one female responds to a particular male. In the following paraphrase from my field notes, females 41 and 23 are interested in male 68. 68 is interested only in 41, with whom he has bred before, and he encourages only her.

> 15JUN82 1330 68 flies to the low platform in the tiny fiddleleaf. He spreads and fans his white tail, then treads. 1331 68 fans his white tail again, then bows, then treads. 1332 68 preens himself. 1333 68 treads, then fans his open white tail. 1335 Female 23, who is molting heavily, looks at 68 from the ground, then flies up to the handrail and looks at him. 68 flies down to ground and walks away. (I got the impression she was going to join 68) .... 1432 68 is treading on a platform in the most northwestern ficus. 23 walks back and forth under the nest on the rail, then flies up to another ficus. 68 stands on the platform. 23 moves around restlessly. 16JUN82 1054 68 flies to the tiny ficus' low platform and treads. 41 immediately flies to the ficus. 1056 23 has been making her way slowly towards 68. Now 41 lands at the nest and walks around. 68 is now nodding up and down with his tail spread and chest low. 1057 41 walks onto the platform. Facing each other, 68 and 41 preen each other violently. 41 walks away. 1058 23 lands at the nest.

> 68 pulls out some of her feathers and she flies to the ground. 1112 68 is in the ficus tree in the northwest corner, not near any platform, throwing his head up and gulping. 41 is also in the tree. 1113 41 has approached and stands next to 68 and they preen each other. Then she walks away.

In the next example, male 1 bows at two sites used in the previous breeding cycle. Females 5 and 64 (1's former mate) both show a little interest.

> 16JUN82 1005 1 bows and treads at the site of the former 22/23 nest. Standing a couple of feet away preening herself is female 5 .... 1014 1 walks out of the old 86/59 nest. He stands on the former 22/23 nest and bows towards female 5 and treads. 5 looks in another direction. 1017 5 flies to the central rubber tree. When I come back from seeing where 5 went, 1 is standing where 5 just was. 1018 64 lands several feet above. 1 immediately turns toward her and bows a few times, then walks over and climbs into the old 86/59 nest. 1020 1 watches 64 fly away.

Trouble may begin when the male attracts more than one female. I observed several aggressive encounters between females over males. In most cases, the male seemed to ignore them. Sometimes one female just displaced the other, but at other times the disagreement was more violent. The females threatened each other, pulled feathers, and pecked toes and feet. Male 86 occasioned the most fights. In all three breeding cycles, he attracted several potential mates. When 86 walked on the ground, as many as three females would follow him at once. The source of his allure is unknown; however, his hackle feathers were a little longer than average for a male (Kathleen Brown, personal communication), and he was one of the heaviest Nicobars. Here are two examples of fights between females over male 86:

> 14JUN82 0948 86 bows at 77 several times, then he hops into the nest basket, spreads his white tail wide, and stands with his chest low. 77 flies over to the basket. 0951 66 lands 3 feet away. 77 flies over and displaces her. 0952 77 hops to land next to basket. 66 lands in same place again. 77 flies at her and displaces her again. 0953 77 returns to the screwpine, then flies to the basket.
>
> 23APR82 1040 86 stands next to 66. 59 looks at them, then moves over and stands on the other side of 86, their sides touching. 86 and 59 stand tails up, white tails spread. 66 pulls 86's tail once, then pecks at 59. 59 moves away. 66 stands next to 86, sides touching.

Fights between females over males occurred in past years as well (Kathleen Brown, personal communication). Although I saw occasionally what appeared to be competition between males over nest sites, only once did I see a fight between two males over a female:

> 20JUN82 1341 Male 22 lands immediately next to female 64. 64 draws away from him. Male 1 reaches over 64's back and pulls at 22's feathers. Then 64 joins in and she pulls at his feathers too. 22 flies away.

Homosexual courtship, although not common, did occur in 1982 and in previous years. During the second breeding cycle of 1982, male 25 responded to male 86's displays. Their courtship followed the same pattern as female-male courtship. 86 later chose a female and built a nest with her. In 1979 and 1980, Nicobar 16, whose sex was never ascertained, built nests with both male and female partners. Apparently 16 played the female role with one male, both female and male roles with another male, and the male role with a female mate. Female-female courtship was limited to a few instances of a female approaching another female who was bowing, and one instance of two females bowing at each other, then one begging.

The second phase of courtship, bonding, begins after two Nicobars have tentatively paired up. They spend much time sitting at or near a prospective nest site. Sometimes the female begs for cropmilk. This may be followed by copulation. Sometimes they groom each other, but usually they just sit quietly, closer than Nicobars normally tolerate each other. When they fly down to the ground, they may eat peaceably from a food bowl while standing next to each other. (Nicobars usually do not allow others at the food bowl, and when they do, they stand as far apart as possible and keep an eye on each other.) The pair then constructs a nest.

On two occasions during this study, male 86 started a nest with each of two females. This is the first time in the history of the colony that this has happened. In the second cycle, he paired up with both 66 and 59, started a nest with 59, worked one day on a nest with 66, then finished the nest with 59. In the third breeding cycle, 86 paired with female 66 and started a nest with her, then paired up with female 35 and started a nest with her as well. He then brought sticks for both nests. Whether he would have carried through with both is unknown, because at this point both 86 and 66 died. Thus, courtship of one female may not be inhibited by being bonded to or nestbuilding with another female.

As indicated by the frequent use of such words as "usually," "frequently" and "sometimes," courtship among the Nicobar Pigeons is quite variable. Almost no behaviors were restricted to only one sex. Female responses to male displays are varied, and male responses to female approaches varied. Faithfulness to the mate varied from individual to individual. Yet a norm exists, so that one can describe typical courtship, typical male behavior, and typical female behavior.

The most striking result of this study is the discovery of "female competition" and "male choice" in the Nicobar pigeon. The sociobiological literature heavily emphasizes that in vertebrates, males compete for females, while females choose among the males. Wilson (1975:324-325) explains:

> Because of anisogamy, females -- defined as the sex producing the larger gametes -- are virtually assured of finding a mate. The eggs are the limiting resource. Females therefore have more to offer in terms of energetic investment with each act of mating, and they are correspondingly more likely to find a mate. Males invest relatively little with each mating effort, and it is to their advantage to tie up as many of the female investments as they can. This circumstance is reversed only in the exceptional cases where males devote more effort to rearing offspring after birth. Then the females compete for males, in spite of the initial advantage accruing from anisogamy.

(See Petrie [1983a] for an example.)

No evidence exists that the male Nicobar cares more for the young than the female does. Yet, many fights were observed between females over males, but only one between males over a female. Some biologists argue that competition between males over nest sites (of which I may have seen a little) is the same as competition over females, since males must first set up territories before they can breed and those with better territories may attract better mates (Mayr 1972:95; Selander 1972). However, the Bird House free-flight area contained a large surplus of nesting sites; even the most popular sites were not in use each cycle. The Nicobars have very small territories. Even in the wild there should be plenty of sites. The Nicobar pigeon is one of an increasing number of examples that demonstrate mate choice cannot be reduced to the formula "male competition, female choice." Although Trivers (1972) had argued that competition and choosiness would vary according to each sex's relative parental investment, until recently this aspect of sexual selection was ignored. Burley (1977) showed that in monogamous species with biparental care, both sexes should exercise some choice, and that this does occur in the pigeon C. livia (see especially p. 209). Halliday (1983) and Petrie (1983b) have defined other circumstances under which males may choose or females may compete. Examples of both sexes being choosy or competitive can be found among insects (Rutowski 1982; Johnson 1982; Hieber & Cohen 1983), fish (Rowland 1982; Weber & Weber 1976), amphibians (Berven 1981; Verrell 1982), mammals (Bell 1983; Lenington 1983; McClintock et al. 1982), and birds (Downhower & Brown

1981; Lenington 1980).

Many of the Nicobar displays are similar to displays in other species of pigeons. Goodwin (1970:28) describes a "Bowing Display" which is like "Bowing" in the Nicobar, and a "high lifting of the feet" (p. 29) which is comparable to the Nicobar "Treading." Other similar behaviors are Goodwin's "Parading" (p. 32) and my "Lowered-Wing Walking," "Nodding" (p. 32) and "Nodding," "Caressing" (p. 35) and "Allopreening," and "Billing" (p. 36) and "Begging" and feeding. I find in Goodwin (1970) no descriptions of behaviors similar to those I have called "Head Throwing," "White Tail Fanning" or "Ritual Looking."

According to Goodwin (1970:29-30):

> A high lifting of the feet, often with a 'marking time' effect accompanies the bowing display of many species. These are intention-movements of mounting as can be clearly seen when, as sometimes happens, a male giving the bowing display carries straight over to copulation without interposing the usual 'billing' ceremony.

However, I believe that Treading may be derived instead from the movements of packing down the nest. First, treading occurs only at potential nesting sites (thus possibly having the side effect of determining whether adequate support for a nest exists). Second, females sometimes tread at the nest sites too. Also, treading looks much like the movements actually used in building, only more exaggerated, as commonly happens in ritualization. It looks nothing like Nicobar mounting or copulation.

The difference in function between Nodding, Head-Throwing, and Bowing is not yet clear. Sometimes they graded into each other. I have no idea why Ritual Looking occurs; however, it only occurs during the courtship period. White Tail Flapping and Lowered-Wing Walking are somewhat more understandable. The former seems an affiliative gesture, while the latter seems aggressive.

The motivation behind bowing seems aggressive as well as sexual. For example:

> 16JUN82 1500 A Nicobar sits at the 86/66 nest, and two more Nicobars are each about 3 feet away. 1502 One of the Nicobars sitting near the nest is 35. 86 flies up next to her, pulls her feathers, then bows a couple of times at her. Then he walks to his nest.

Frequently an approaching female who is later accepted as a mate is attacked rather than welcomed. This would not be expected if the male's motivation were sexual only. Twice I saw a male bowing at another in what seemed to be a dispute over a nest site. Once I saw a Nicobar male bow after he and the female he was courting were chased off their site by a tern. Also, of the three females I saw bow, females 77 and 35 each bowed after having been rejected by males in which they had shown strong interest. 35 stood near the nest site of the male and his new female while bowing; 77 stood by the site she and the male had chosen, but bowed in the direction of the male's nest with his new female. Goodwin (1970:28) says that bowing:

> is primarily a sexual display and is shown at fullest intensity by a male towards females to whom he is strongly sexually attracted. In many species, however, it is also used in threatening or defensive contexts.

Fabricius & Jansson (1963:545), who studied rock doves (_C. livia_), reported that "[b]owing is primarily an intimidating display," and that it

> seems to have the double function of intimidating strange males intruding into the territory, and of activating the sexual instinct of the female.

Male displays attract females in other pigeon species as well. Frith (1977:169-170) described male bobbing displays that attracted sexually active females to fly up and join the males in the pigeon species _Chalcophaps indica_, _Leucosaria melanoleuca_, and _Geophaps scripta_. Sometimes among feral pigeons, _C. livia_, males take possession of nest boxes or ledges and produce calls that attract unmated females to join them (Burley 1977:47).

I suspect that the Nicobar female is attracted primarily to the male, and secondarily, if at all, to his nest site. The same females tended to approach a male no matter where he displayed, and pairs did not break up if the nest were constructed at a different site than where pairing occurred. Also, when the Nicobars choose the same mate again they frequently build at a new location. Chapter 4 will further develop the point that females are attracted to males, not sites.

Summary

Courtship begins when a male chooses one or more potential nest sites and begins bowing and treading on them. Females are attracted by him and his display. When one approaches, he may chase her away, or they may preen each other. When more than one female is interested in a male, they may fight over him. After he has chosen one female, then they spend some time before beginning their nest sitting close together, grooming each other, and copulating. Courtship among these birds is flexible and deviations may occur, such as females bowing, males begging, and homosexual courtship.

Chapter Four

Mate Selection

During courtship the Nicobars make choices. Females must decide which males to approach, and males must choose a mate from among the females who visit his site. In this chapter I describe the Nicobar tendency to choose mates of the same age.

According to Burley (1981:515), one can only discover mate preferences experimentally (for example, Burley 1977) and not through observations of naturally occuring choices. In species in which both sexes invest in young, the most attractive animals will probably woo and win attractive mates. But the unattractive animals may not be able to find a mate at all, or may have to settle for an unattractive mate. In the words of the popular song, "If you can't be with the one you love, love the one you're with" (Stills 1970). The choices of unattractive birds thus do not reflect their preferences.

Many species show patterns of mate choice, such as assortative mating. In human societies, people pick mates of a similar age. Europeans and Americans also select those similar in intelligence, height, hair color, and eye color (Damon 1977:108). The brentid weevil (*Brentus anchorago*) mates assortatively for size (Johnson 1982). Black-legged Kittiwake gulls (*Rissa tridactyla*) choose mates of similar reproductive experience (Coulson 1966). White-crowned sparrows (*Zonotrichia leucophrys*) seem to mate assortatively for song dialect (Baker & Mewaldt 1978; Baker 1982), and mourning doves

(*Zenaidura macroura*) for dominance rank (Goforth & Baskett 1971). Midas cichlids (*Cichlasoma citrinellum*) and lesser snow geese (*Anser caerulescens*) choose mates of the same color (Barlow *et al*. 1977; Cooke *et al*. 1976). Wood lice (*Porcellio scaber*) mate assortatively for an enzyme polymorphism (Sassaman 1978).

Burley (1977:238) suggests that assortative mating cannot be assumed to be due to animals preferring similar mates. Rather, when both sexes prefer the same qualities in the opposite sex and when these qualities occur in the same proportions in males and females, assortative mating may reflect only the ability of attractive animals to win attractive mates, leaving the unattractive animals with each other. Thus, although studies of natural mate choices are important for finding out what patterns of choice actually occur, they cannot tell us the animals' preferences.

Between April 1976 and December 1982, 13 female and 12 male Nicobars bred at the Lincoln Park Zoo. 83 pairings produced 85 eggs, 80 of which had been officially recorded and five of which I discovered through examination of keeper's logs, records of eggs, and miscellaneous zoo records.

I originally collated this data to search for effects of mate choice on breeding success and vice versa. Birds that paired up but did not lay an egg are not included in this analysis. Birds are not considered potential mates until they have bred once; others are left out to avoid the problem of deciding which non-breeding birds were reproductively mature. The two cases of twin eggs are counted as two pairings or four eggs, depending on the analysis. In two cases, only one parent of an egg was known. These cases are left out of the analyses in which knowing the identity of both parents is required.

I found that neither breeding success nor failure influenced choice of mate in this colony. Both females (Table 4-1) and males were as likely to retain their mates after failure as after success. Of 68 total transitions between mates for females between 1976 and 1982, 40 (59%) were to the same mate. Of 69 transitions for males, 35 (51%) were to the same female.

An important influence on mate choice was age. The Nicobars tended to choose mates of the same age. In Figure 4-1, all pairings between Nicobars of known ages is shown. The correlation coefficient for ages of mates is .55. Because the ages of the birds were not bivariate normally distributed, I used Spearman's Rank Correlation Coefficient to test for significance. R_S = .52, which was significant at the .01 level.

Figure 4-2 shows the relative ages of mates for all pairings. Although the ages of the original colony founders are unknown, none of these birds hatched later than 1975. Thus they are all older than the captive hatched, who hatched in 1976 and later. These older birds who were left out of the scattergram in Figure 4-1 also show a strong tendency to mate with each other rather than with the younger birds. A secondary trend for older, presumably wild-caught males and younger females to mate is also evident.

Despite the obvious pattern, Figure 4-2 still contains 32 exceptions to the finding that Nicobars mate with others of the same age. Figure 4-3 shows the same graph after removal of pairs in which at least one of the members had no age-mate of the opposite sex. It was impossible for these birds to choose a mate of the same age. These cases account for 47% of the exceptions.

Another potential difficulty for birds preferring to mate with others of the same age is that some age cohorts contained only one member of one sex and two or more members of the other sex. Since pigeons are monogamous, some birds in this situation must either not mate or choose a mate of a different age. Figure 4-4 shows that the remaining exceptions disappear when one removes those cases in which one member of the pair belonged to an age cohort with an unfavorable sex ratio.

Thus all exceptional pairs contained at least one member for whom a mate of the same age was either completely unavailable or much competed for. The motivation of the other member of such pairs is puzzling. For some, mates of the same age were available, and yet they chose to mate with a different aged partner.

For each year I tested whether or not the proportion of the mates who were the same age was higher than that expected by chance, using a binomial distribution (Freund & Walpole 1980:403-404). Original colony founders were assumed to be all the same age for this one-tailed test. Since the first captive-hatched males did not reach breeding age until 1979, earlier years are left out. Although the proportion of same-age mates was not greater than expected by chance in 1979 ($P < .07$), birds of the same age did pair together significantly more often in 1980 ($P < .01$), 1981 ($P < .05$), and 1982 ($P < .0001$).

Table 4-1

Effect of Success on Mate Retention

(female perspective)

	T_2	
T_1	keep	change
successful	18	11
unsuccessful	20	11
total	38	22

chi-squared = 0.0479, 1 df, NS

Changes due to absence of former mate are left out (N=5).
Successful adoptions (N=2) count as successes.

FIGURE 4-1

RELATIVE AGES OF MATES

(based on 31 pairs of biras of known age)

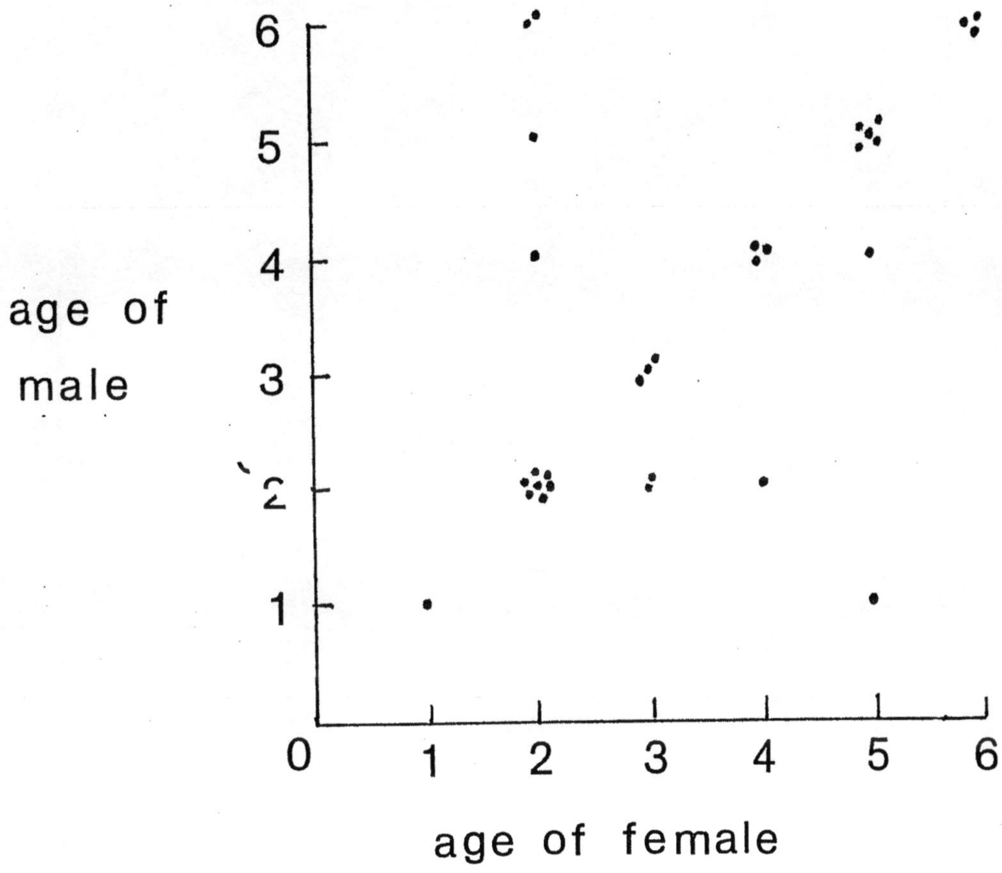

$R_S = 0.52$

FIGURE 4-2

RELATIVE AGES OF ALL MATES

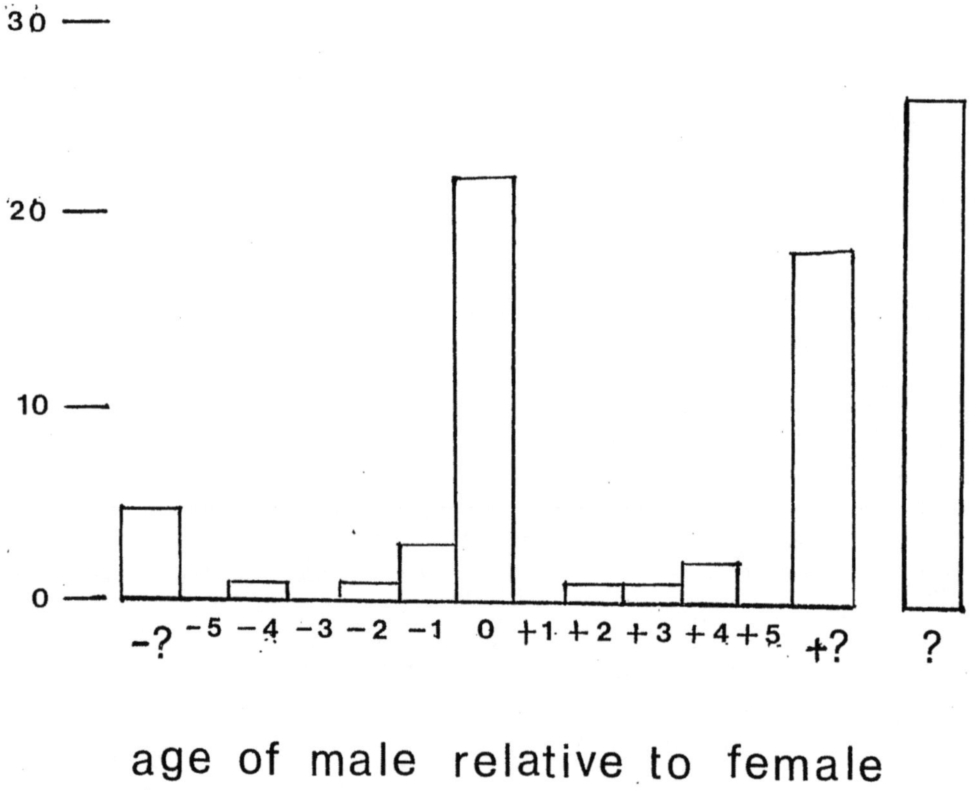

FIGURE 4-3

RELATIVE AGES OF MATES WITH AGE-MATES OF THE OPPOSITE SEX

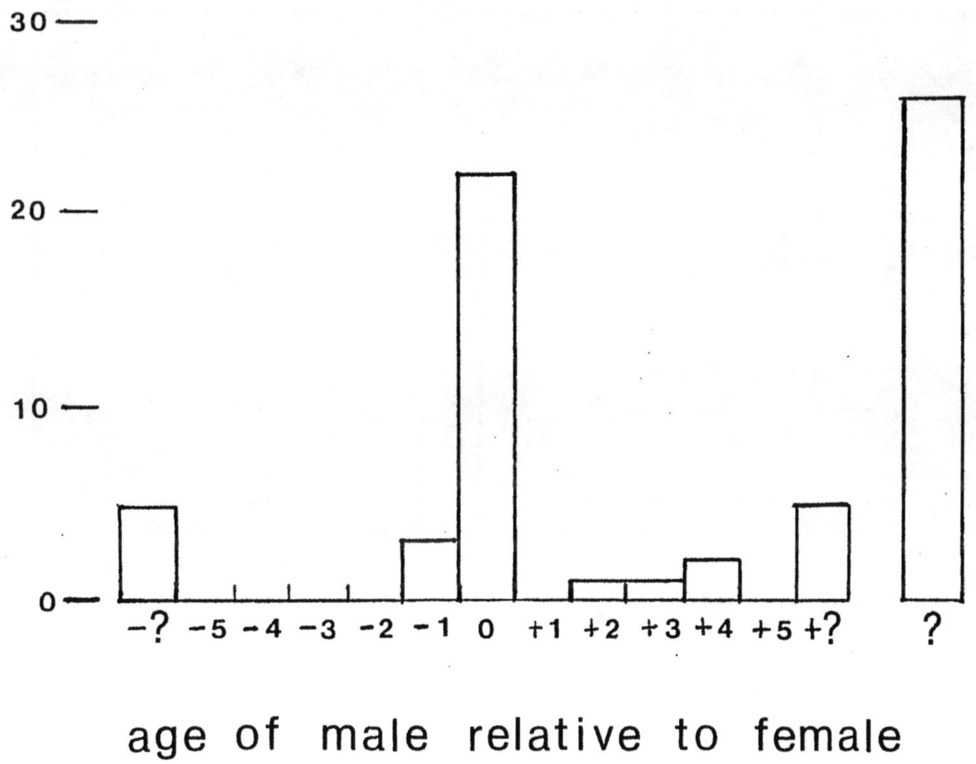

FIGURE 4-4

RELATIVE AGES OF MATES WHO BE-LONGED TO AN AGE COHORT WITHOUT A SKEWED SEX RATIO

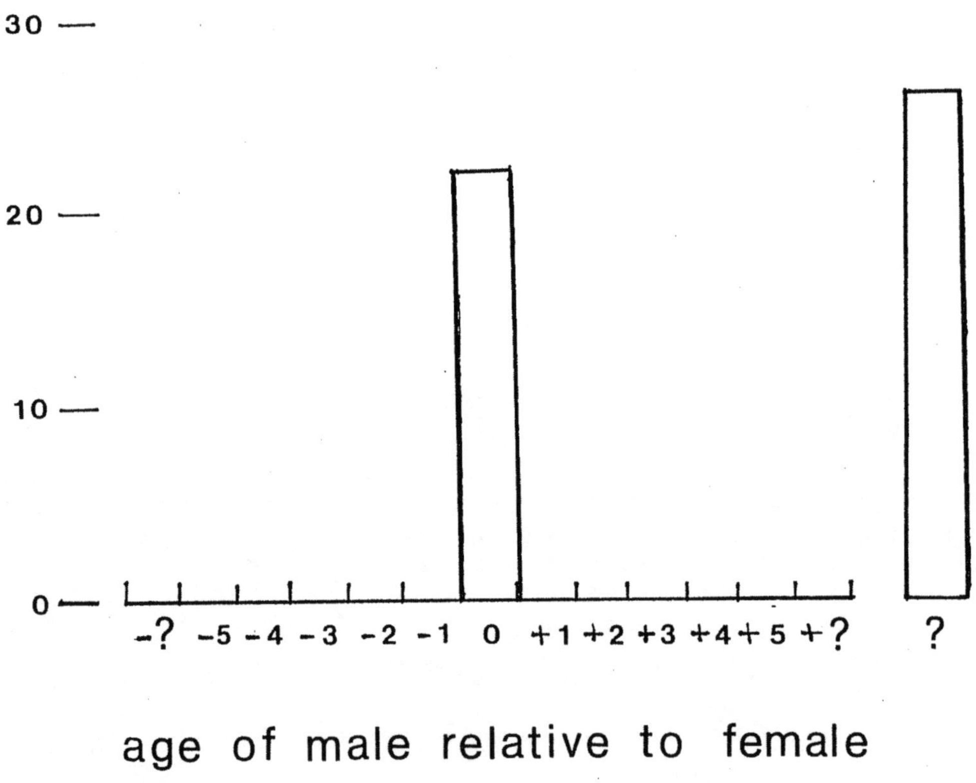

Frequently the effect of assortative mating on a population resembles that of inbreeding (Wallace 1968:120-121). When the favored traits have a genetic basis, homozygosity may increase in the population. This happens when the features that the partners share and for which they have chosen each other are caused by the same alleles. Thus mates resembling each other phenotypically may resemble each other genetically as well, as if they were related. Also, matings between relatives become more frequent when animals mate assortatively since two relatives are more likely to resemble each other than two non-relatives are.

However, I believe that in the Nicobar pigeon assortative mating instead reduces inbreeding. If a Nicobar pigeon chooses a mate from its own age cohort, it is automatically prevented from mating with its parents, grandparents, children, and grandchildren. It is also usually prevented from mating with siblings. Females laid between 0 and 4 clutches per year (M = 2.03, SD = 1.13) and males fathered between 0 and 5 (M = 2.14, SD = 1.13). However, the average number of chicks fledged per year for members of both sexes was 1.00 (SD for females was .72, SD for males .83). Females only fledged two chicks per year 27% of the time. The chance then of a Nicobar having a half or full sibling of the opposite sex in its age cohort is not very good. (One might expect an even smaller chance under natural conditions, given the more rigorous environment and possibly shorter breeding season.) If a Nicobar does have a relative in its age cohort, the relationship is likely to be more distant than that of parent-child or siblings. Thus matings between birds of the same age should produce young that are less inbred.

I tested this expectation both theoretically and using the actual data. First, for each year I calculated the expected average inbreeding coefficient, $\bar{F}$, for the offspring if (a) the breeding birds mated at random, and (b) if they mated exclusively with those of the same age. During 1977 and 1978, inbreeding would not have occurred if mating was only between same-aged birds. During 1979, 1980, 1981, and 1982, random mating would have produced young 3.4 to 7.9 times as inbred as strict assortative mating by age. The $\bar{F}$'s expected of offspring from random mating for those years ranged from .025 to .064, while the averages of the F's expected from assortative mating were only .007 to .017. Using a one-tailed Z test, I tested whether the average inbreeding coefficient for assortative mating was significantly smaller than that for random mating in 1979, 1980, 1981, and 1982. The $\bar{F}$ for assortative mating was significantly lower for the 1979 data ($Z = 2.75$, $P < .005$), as well as that for 1980 ($Z = 2.65$, $P < .005$) and for 1982 ($Z = 3.75$, $P < .001$). The $\bar{F}$'s for 1981 were not significantly different ($Z = .57$, $P < .28$). These results should be viewed with caution, however, because the sample sizes from the assortative mating results were small.

The qualitative prediction that mates of the same age would produce less inbred young proved correct not only theoretically, as shown above, but in reality as well. The $\bar{F}$ of eggs produced from different aged parents was .0508, while $\bar{F}$ of eggs of same-aged parents was just .00625, an eight-fold difference. Even if matings between colony founders (all counted as same-age matings between non-relatives) are left out, $\bar{F}$ for same-aged mates only rises to .014 -- still a 3-1/2 fold difference. Higher inbreeding between birds that do not mate assortatively for age is also seen if one examines only the eggs produced by incestuous matings (between relatives of any degree). The five from same-aged mates had an average inbreeding coefficient of .0625, while the other ten eggs had an $\bar{F}$ of .1625, an average 2.6 times as high.

Incest avoidance is not only a human trait. Mechanisms in the societies of other vertebrates may also prevent incest (see Bischof [1972] for a discussion of potential mechanisms, and Packer [1979], Pusey [1980], Grau [1982], and Hoogland [1982] for examples). These vertebrates include macaques, chimpanzees, geese, ducks, rats, and prairie deermice (Parker 1976:290-291). Both in humans and in these other vertebrates, "long-term and early association seem to be the crucial variables" (Parker 1976:297) in causing a prospective pair to find each other sexually unattractive.

In humans inbreeding leads to lower birth weights and to increased miscarriage, stillbirth, and infant death rates (Levitan & Montagu 1971:520-521). Studies on other animals also demonstrate inbreeding depression. Wright (1977:44-93) summarized a large literature on captive animals, showing decreased litter sizes, decreased percentage of live young, and/or decreased viability of young in guinea pigs, rats, mice, rabbits, pigs, sheep, poultry, quail, and fruitflies. Cothran _et al_. (1983) showed that homozygosity in wild deer was associated with lower fetal weight. In mosquitoes, _Culex pipiens_, higher inbreeding led to fewer eggs, lower fertility, and higher larval mortality (Boesiger 1982). Inbreeding depression has also been shown in natural populations of great tits, *Parus major* (Greenwood *et al*. 1978; Noordwijk & Scharloo 1981), and of baboons, *Papio anubis* (Packer 1978), and in several species living in zoos (Ralls *et al*. 1979; Roychoudhury & Sankhala 1979).

Recently inbreeding and outbreeding have been viewed as having both good and bad effects (Shields 1982; Partridge 1983:235-239; Bateson 1983:259). Noordwijk and Scharloo (1982) found that the lowered hatchability of inbred clutches of tits was compensated for by the higher probability of those that did hatch surviving to adulthood. However, the rarity of highly inbred species indicates that the bad consequences of inbreeding usually outweigh the good (Partridge 1983:239).

Avoidance of inbreeding may be especially critical in the natural history of the Nicobar pigeon. I discussed in Chapter 1 the long migrations made by Nicobars. These founding populations must frequently be small, and thus subject to Founder's Effect. Denniston (1978), for example, showed that although a sample of ten individuals drawn at random from a larger population retains 95% of the genetic variance found in the parent population, only 13 generations are needed to reduce the variance to 51% of that in the original population. Rare genes (which contribute little numerically to genetic variance) are frequently lost much faster (Denniston 1978). Inbreeding would hasten this already rapid diminution of the gene pool. A mechanism to retain heterozygosity in the Nicobar pigeon as long as possible would be quite valuable.

Whether assortative mating by age also occurs in the wild Nicobar pigeon is completely unknown. (In Chapter 2, I discussed the close similarities between captive and wild Nicobar behavior). However, the Nicobar pigeon, despite its wide distribution over small islands, can only be broken down into two subspecies (Goodwin 1970:263). This indicates that the wild Nicobars have maintained a gene pool that is phenotypically homogenous over vast distances despite the physical isolation of local populations. If the Nicobars reduced inbreeding through assortative mating by age, genetic drift would be retarded in local populations, providing a solution for this puzzle.

How do Nicobars recognize others of the same age? Any phenotypic clues are indiscernable to human eyes. Feral pigeons, C. livia, of different ages also look alike to us, but can be distinguished by conspecifics (Burley 1977:22, 160). The sensory capabilities of C. livia greatly exceed our own (Kreithen 1979; Delius & Emmerton 1979). Therefore the possibility exists that the Nicobars can judge each other's ages.

"Long-term and early association" (Parker 1978) with an individual does not seem to lead to avoidance of that bird as a mate. Nicobars who did not mate with others of the same age sometimes mated with their mothers or fathers, who were their only associates when in the nest.

If the Nicobars cannot judge the age of other Nicobars, one possibility is that same-aged pairs are brought together both passively (by being rejected as mates by birds of other ages) and actively (by selecting more familiar birds, such as those who fledged near the same time). This could happen because the Nicobars frequently keep the same mates. The young bird breeding for the first time might find as the only unpartnered birds those also breeding for the first time. It in turn, having chosen a partner of the same age, might return to it again in the future in preference to birds of other ages. Each age cohort is shunned by other birds, and it in turn shuns the younger ones. Such a system would be expected to break down when a bird is unable to find a mate in its age cohort. And, as I showed earlier, this does happen in the Nicobar.

Preference for former mates cannot be the total explanation, however, because then a bird changing mates should show no preference for same-aged versus different-aged birds. Male 68, however, mated almost entirely with the two females in his age cohort. Female 35 mated exclusively with three of the four males in her cohort, until all had either died or been sent to other zoos.

In 1982 I studied chick rearing. After fledging, the two chicks spent a considerable amount of time together on the ground, and they interacted with the third youngest Nicobar more than with any of the older birds (see Chapter 7). 22 and 23, which had bred together at another zoo, came to Lincoln Park in 1980, and bred with each other six times before 22 switched. The two Lincoln Park females he then mated with had both hatched the year he arrived. 23 has not found any new mates (as of December 1983). Two wild-caught birds, K29 and 130, arrived together in 1976 and mated with each other twice before mating with any of the other birds. Four birds arrived at Lincoln Park unpaired (24, 25 (LCB 80-35), 25 (LCB 79-25), and 43). 43 did not breed until his third breeding season at the Zoo, 24 and 25 (LCB 79-25) each spent one breeding at the Zoo and did not breed, and 25 (LCB 80-35) who arrived as a juvenile has not bred as of his fourth season. This anecdotal evidence suggests that chicks fledging about the same time associate with each other and become familiar with each other, and that familiarity, far from discouraging mating, actually encourages it. The abstinence of new birds in the colony is difficult to explain otherwise.

Summary

Mate choice was not random. Rather, individuals showed a statistically significant tendency to mate with others of the same age. Half the exceptions involved birds which, having no age-mates, mated with birds of a different age. Assortative mating in this colony reduces inbreeding. Prevention of inbreeding may be important in the natural history of the Nicobar pigeon, given its migratory habits. Founding populations are likely to be small and thus susceptible to the negative effects of inbreeding.

Chapter Five

Nest Building and Choice of Sticks

After finishing courtship the Nicobars build nests. As in other pigeons (Goodwin 1970:44), the male usually collects the sticks and the female usually builds them into a nest. The female is thus alone at the nest during much of the construction and the weight of defense falls upon her (Chapter 6). A sequence analysis of the construction behaviors is planned for the future; in this chapter I will briefly describe nestbuilding, and then concentrate on analyzing an experiment on stick preference.

Construction techniques varied greatly not only from pair to pair, but even in the same individual over time. The shortest construction time in 1982 for a nest was only three to four days, the longest, eight to nine days. Usually the Nicobars built on wire mesh platforms fastened into trees by the keepers; however, sometimes they chose forks of trees or even built precariously on branches with very little support. As might be expected from colonial birds, sometimes a tree would contain more than one nest, sometimes only a few feet apart. A pair's building might be successful right away, or a pair could work vigorously for two days with only a pile of fallen sticks under the nest site to show for it. Sometimes my daytime observations seemed to encompass most of the construction of a nest; other times a pair rarely worked on its nest during the day, but each morning would be found sitting on an enlarged nest. The fastest stick collection trip by a male took only 1 minute 39 seconds, although sometimes males did not return to their nests for several hours.

In many bird species, the existence of a nest or possibly the experience of building a nest seems necessary for the female to lay an egg (Lehrman 1961:1280-1281; Lehrman 1964:154-155). In contrast, in 1982, a female Nicobar without a mate or nest laid a fertile egg. Female 77 had courted with male 86, who chose another female instead. Later I observed her approach 22's nest (on which sat his mate 59) and copulate with him. Four days later she began sitting in a breadbasket hung up to replace a destroyed touraco nest, and the next day was discovered to be sitting on a fertile egg. The egg was not pulled, and 77 looked thinner and dirtier every day. When she finally deserted it after 13 days, the egg was put into the incubator but died.

From 17 April to 17 May 1982, I conducted an experiment to discover the Nicobars' preferences for nesting material. My questions were:

1. Do the Nicobars prefer limp or rigid materials?
2. Do the Nicobars prefer green or brown materials?
3. Do the Nicobars prefer materials 6, 12, or 18 inches long?
4. Do the Nicobars prefer materials with projections or without?

To answer these questions, I first removed all sticks from the aviary floor, and then collected, cut, and put out 1903 pieces of potential nest material. To answer question 1, I put out both sticks and rug yarn. To answer question 2, the rug yarn was either green (Woolco rug yarn color 671, Avocado) or brown (Woolco rug yarn color Dark Brown). To answer question 3, 6-inch, 12-inch, and 18-inch long sticks and pieces of each color yarn were put out. To answer question 4, the sticks were cut to be either "straight" (not literally straight, but rather having no projections) or "forked" (having at least one projection at least one inch long). Due to the huge number of sticks to be collected and cut in a short time, I did not measure their thicknesses. To try to control for any preference by the Nicobars for particular thicknesses, I did try to have a range of thicknesses, and to have similar thicknesses in all the categories. I put out 600 pieces of yarn on 17 April 1982: 200 strands each of 6-inch, 12-inch, and 18-inch yarn. One hundred of each length were green, the other hundred brown. Table 5-1 shows the dates and numbers of sticks put into the aviary.

The first two days of collecting I tried to gather as many sticks as possible. Because straight sticks were more common than forked in the Zoo's Rookery, straight sticks were over-represented at first. Then, because forked sticks disappeared so quickly (partially due to heavy collecting by white-headed buffalo weavers, Dinemellia dinemelli), I concentrated on putting out as many forked sticks as possible. By the end of the experiment more forked sticks were in the aviary than straight.

In gathering the data, I concentrated primarily on one nest and opportunistically noted males taking sticks to other nests. Chicks were fledging towards the end of the experiment, diverting my attention from the end of nest building. Thus the following numbers cannot be used to compute rates of building or to compare enthusiasm of different males.

Courtship began 15 April 1982, and five pairs built nests during the experiment (Table 5-2).

The experiments showed that the Nicobars prefer rigid materials to limp. No yarn was used in any of the nests. However, the Nicobars were observed to take 43 pieces of "vegetation" (dried or fresh leaves, stems, grass, or other soft plant materials) to the nests, and some were noticeably limp. They may have rejected the yarn due to some property other than limpness. When I ended the experiment 17 May 1982 and attempted to collect and remove the yarn, I found none. I have no idea how 600 pieces of yarn disappeared. The Nicobars did not collect the yarn while it was still there; that the yarn disappeared is regrettable but I doubt it influenced the experiment.

Because the Nicobars showed no interest in the yarn, I was unable to determine their relative preference for green versus brown materials.

The Nicobars strongly preferred longer sticks (12 and 18-inch) to short sticks (6 inches). (12 and 18-inch sticks will be considered together in this chapter. 12-inch sticks are 100% longer than 6-inch sticks, but 18-inch sticks only 50% longer than 12-inch. I discovered I could easily make the first distinction when observing, but frequently confused the second.) 348 pieces of material could be identified as being 6-inch (112 sticks, 32%) or 12 and 18-inch (236 sticks, 68%). If the birds chose randomly, one would expect them to choose 41% 6-inch sticks (536 6-inch sticks put out/1303 total sticks put out). A two-tailed Z test of the difference of proportions showed statistically fewer 6-inch sticks used than expected by chance ($Z = -3.40$, $P < .001$).

The Nicobars also strongly preferred forked sticks to straight. I observed 289 forked (76%) and 89 (24%) straight sticks taken to the nests. The expected proportion of forked sticks if the birds chose randomly was 694/1303, or .53. A two-tailed Z test of the difference of these proportions showed statistically more forked sticks were used than expected by chance ($Z = 9.04$, $P < .0001$).

For all but 11 cases, I identified the sex of the bird bringing the stick. Analysis showed that females differed from males in three ways: quantity of sticks, types of sticks, and timing of sticks. First, males brought 414 sticks to nests during my observations; females brought only 37. Thus males were the primary collectors.

Second, using a binomial test (Freund & Walpole 1980:403-404), I found that females brought significantly fewer forked sticks and more straight sticks than the group as a whole ($P < .007$). Females did not differ significantly from the group in proportions of 6-inch and longer sticks brought ($P < .09$). I have already shown that the shapes and lengths of sticks brought to the nests differ significantly from the proportions I put out. However, sticks chosen by females alone do <u>not</u> differ. Using the binomial test again, I compared the proportions of sticks chosen by females to the proportions of sticks occurring in the aviary. The females seemed to be choosing randomly; H_0 could not be rejected for either the proportions of forked and straights ($X = 13$; critical region: $X \leq 7$ or $X \geq 17$) or of the proportions of short and longs ($X = 13$; critical region: $X \leq 6$ or $X \geq 16$).

Third, males brought most (96.8%) of their sticks before the egg was laid. Females brought most (89.2%) of their sticks after the egg was laid (Table 5-3). This difference was statistically significant (chi-squared = 263.32, 1 df, $P < .0001$, phi = .77).

Nestbuilding itself can be broken down into two parts, each lasting two to four days. To divide the data, I counted the days each pair worked on their nest before the egg was laid, and split it at the median. Thus nests assigned to the same group may have been at different stages of completion. The Nicobars do not choose the same types of sticks throughout nestbuilding. Compared to the first half, straight sticks and vegetation are more frequently chosen in the second half, and forked sticks, less frequently (Table 5-4). This difference was statistically significant (chi-squared = 48.65, 2 df, $P < .001$, phi = .36). Also, the proportion of 6-inch sticks went up in the second half, and the proportion of 12-inch and longer sticks went down (Table 5-5). This difference is also statistically significant (chi-squared = 10.26, 1 df, $P < .005$, phi = .18).

Before discussing possible meanings and implications of these results, I should discuss their accuracy. The preceding analyses assume that non-Nicobars removed sticks randomly. Although I made no systematic observations of their preferences, I did notice that forked sticks were more heavily collected by non-Nicobars than straight. This only strengthens the previous results regarding forked sticks. Also, these sticks are not a perfect representation of the sticks in the nests. Nestbuilding took place at night and while I ate lunch, so that I was unable to observe all the sticks added, even to the primary nests. Perhaps more important is that not all sticks remained in the nest. For example, on 23 and 24 April 1982, I saw 68 and 41 bring 53 sticks to their nest site. Each stick fell out of the tree soon after. 68 began collecting sticks from the pile that formed under the nest, and I counted some sticks several times, as each was collected and taken to the nest again. These results thus are more a reflection of the types of sticks preferred (a pigeon can cast multiple "votes" for the same stick), not the types of sticks used (a stick can only be represented in the nest once).

On 12 May 1982 I collected the deserted 1/64 nest. During the experiment I had observed male 1 take 114 pieces of material to this nest. Five I was unable to identify, 8 were pieces of vegetation and 101 were sticks (90 forked, 11 straight). When I dissected the nest, only 80 sticks and 4 pieces of vegetation were in it. 70 sticks were forked and 10 straight. I tested the difference of the proportions of forked sticks observed taken to the nest and of forked sticks actually in the nest using a two-tailed Z test, and found it not to be significant ($Z = .4894$, NS). No other nest built during the stick experiment could be collected; based on this single nest, the stick experiment data seems to be a reasonable approximation of sticks actually in the nest. Thicknesses were not measured, since I had no data on thicknesses of sticks observed taken to the nest.

I designed this experiment with the assumption that, to the Nicobar pigeon, not all sticks are equally valued. I further assumed that preferences would sometimes reflect the stick's quality as a nest ingredient. I learned when I cut the sticks into various lengths and put them into piles that the sticks in the straight piles could be easily separated, while those in the forked piles became entangled immediately and were hard to separate. The pigeons chose preferentially materials that were rigid and forked; such materials would contribute to the cohesion of the nest by resisting forces (wind, nest raiders, the parent's own activity) tending to pull it apart.

The Nicobars also chose the longer sticks. Such sticks may also help hold the nest together better than short sticks. The longer the stick, the more sticks with which it can be interwoven. Upper bounds on length are set not only by strength of the individual, but also by the diameter of space it must fly through to reach its nest.

The increase in short sticks, straight sticks, and vegetation in the second half of nestbuilding suggests that the Nicobar pigeon nest contains a rudimentary lining. The long forked sticks already having provided a secure base, the Nicobars then bring smaller and limper materials. These materials can be woven together more tightly and smoothly than the longer sticks with their many projections, providing better insulation and comfort. This lining is more apparent statistically than visually. I had looked for linings and not seen them. No one else seems to have noticed them either: Goodwin (1970:263) says the Nicobar possesses the "usual pigeon nest," and Baker (1928:214) says, "nests are the usual badly put together fragile platforms of sticks," and (1913:115), "The nests were ... regular Pigeons' nests, merely a platform of twigs, very loosely and carelessly put together and without lining of any kind." The Lincoln Park Nicobar nests look like this as well. This experiment shows that the nests have a structure despite their appearance to the contrary.

Further, although less clear-cut, results were obtained. Although most nests were built too high to retrieve, I was lucky enough to collect three, all built on the same platform. The K22/K24 nest had been deserted in April 1982 after their newly hatched chick died. 1 and 64 deserted their first attempt at nestbuilding and rebuilt in another tree. 86 and 35's incomplete nest was deserted by 35 in June 1982 after 86 died. Sticks were lost from all three nests during collection. I removed the sticks from top to bottom, hoping to approximate in reverse order the sequence in which they had been laid down. I recorded the length of every item over three inches long, and noted whether it was a stick or vegetation, and whether it was forked or straight.

Figure 5-1 shows a smoothed curve of the lengths of the sticks from top to bottom. None of the three nests show an increase of short sticks at the top, as my observations showed. (Dips at the end of each graph are probably due to broken pieces of stick falling to the bottom of the storage bags.) My observations of nestbuilding also revealed that more vegetation was collected in the second half of building. Vegetation was scattered evenly through the 1/64 and 86/35 nests, but was concentrated in the top of the K22/K24 nest, with 64% of the vegetation removed among the first 50% of materials.

Table 5-1

Sticks put into the Aviary

Date	6" straight	12" straight	18" straight	6" forked	12" forked	18" forked
4/17/82	170	142	95	64	85	42
4/18/82	48	9	0	3	0	1
4/20/82	0	0	0	43	9	5
4/22/82	6	17	4	26	44	20
4/27/82	7	18	21	50	75	13
5/2/82	6	8	1	52	29	4
5/4/82	16	21	20	45	68	16
Totals:	253	215	141	283	310	101

Table 5-2

Pairs Building Nests During Experiment

Pair	Start Date	End Date	Number of Sticks Observed
K22/K24	4/22/82	egg laid 4/26/82	43
68/41	4/23/82	egg laid 4/29/82	212
86/59	4/29 or 4/30/82	egg laid 5/3/82	69
86/66	5/2/82	5/2/82	3*
1/64	5/2/82	deserted 5/9/82	117
1/64	5/10/ or 5/11/82	egg laid 5/17/82	16

* These three sticks are added to the 69 in the 86/59 nest for analysis.

Table 5-3

Effect of Sex and Presence of Egg on Number of Sticks Collected

	Females	Males
Sticks collected during nestbuilding	4	399
Sticks collected on day of egg laying or after	33	13
Total	37	412

Table 5-4

Effect of Stage of Nestbuilding
on Shapes of Sticks Taken to Nests

	Straight Sticks	Forked Sticks	Vegetation
First Half of Nestbuilding	23	188	11
Second Half of Nestbuilding	52	80	23

Table 5-5

Effect of Stage of Nestbuilding on Lengths of Sticks Taken to Nests

	6" Sticks	12" & 18" Sticks
First Half of Nestbuilding	48	149
Second Half of Nestbuilding	49	69

FIGURE 5-1

LENGTHS OF STICKS REMOVED FROM NICOBAR PIGEON NESTS

from top to bottom

(smoothed 3RSRH3 , Tukey 1977:210-234)

x axis is stick number

y axis is stick length

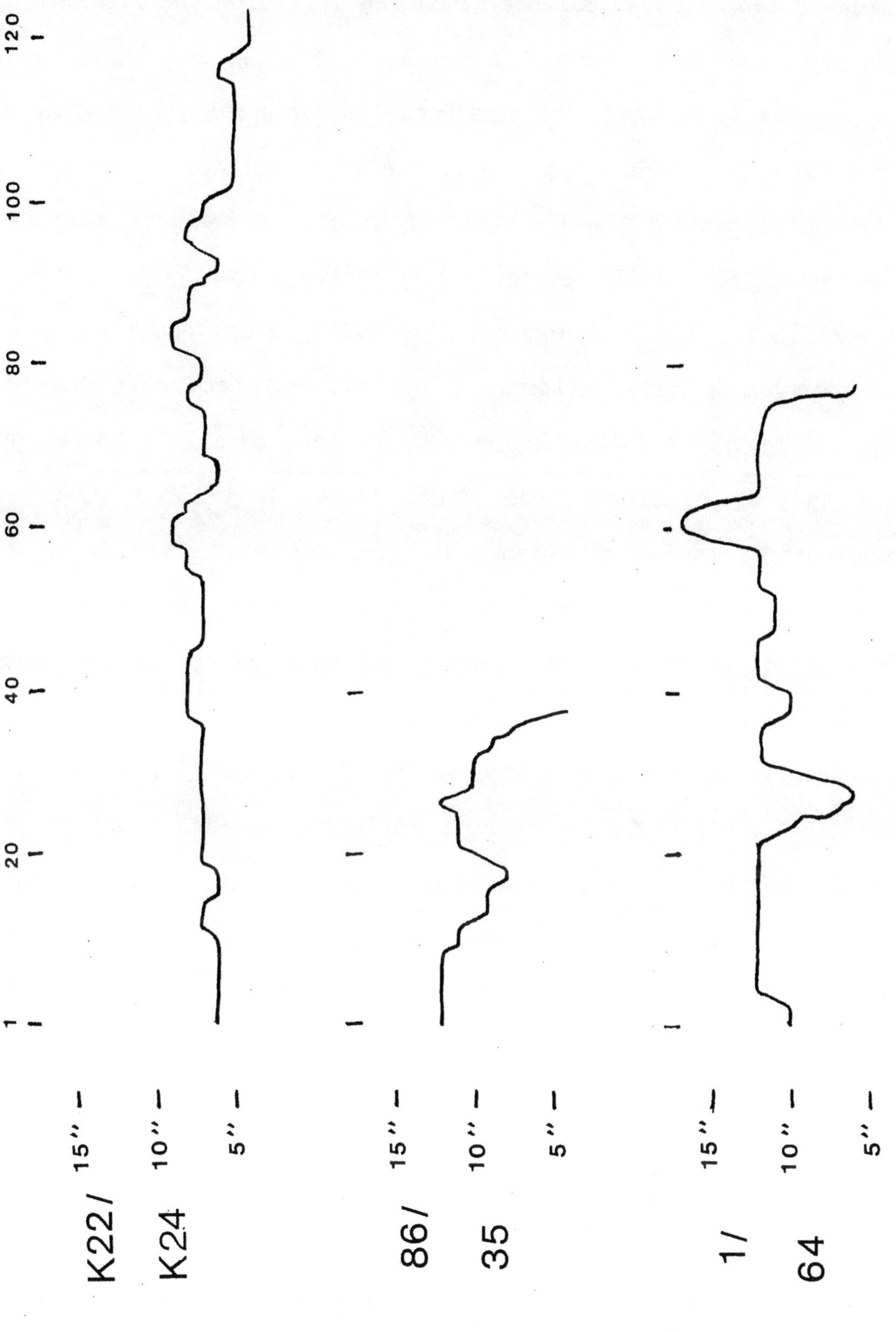

K22/
K24
86/
35
1/
64
15"
10"
5"
1
20
40
60
80
100
120

Plate 5-2 shows the neat, compact nest of K22 and K24, each a veteran of over 15 previous nests, on the left. On the right is the loose and bulky first nest of 1 and 64. Both had been built on the same platform. The reader is cautioned against jumping to the obvious conclusion -- that the difference is due to experience. Two-tailed Z tests showed that the mean length of sticks in the 1/64 nest (M = 10.94 inches, SD = 3.67) was not significantly different from those in the 86/35 nest (M = 9.92 inches, SD = 2.92), even though 86 and 35 were experienced nest builders (Z = 1.61, NS), and that the mean length of length of sticks in the K22/K24 nest (M = 7.10 inches, not including over 50 pieces under three inches long, SD = 2.36) was significantly shorter than those in the 1/64 nest (Z = 8.30, P < .0001). Z-tests of the difference of proportions of the types of sticks chosen by inexperienced male 1 and those chosen by the other, experienced males in the stick experiment showed the only significant difference to be that 1 used more of the 12-inch and 18-inch sticks (Z = 2.12, P < .05). K22, the oldest male in the experiment and the only wild-hatched, built a nest even more divergent from the other nests. The materials K22 collected were significantly less often forked (Z-test of proportions, Z = -9.38, P < .0001), were more often straight (Z = 4.87, P < .001), and were more frequently vegetation (Z = 7.27, P < .0001). He also used more 6-inch and fewer 12 and 18-inch sticks than the others, but not significantly so (Z = .99, NS). The individual differences among males seem too great to draw any conclusions from this data about the effect of experience on nest building, but could be explored further in more refined experiments.

Plate 5-2

Nests of K22 and K24, and 1 and 64

Summary

To determine Nicobar stick preference, I distributed three lengths of yarn in two colors, and three lengths of sticks, some with projections and some without. No Nicobar used yarn in its nest, although the birds did find and use limp pieces of vegetation. The Nicobars preferred the longer sticks to the shortest ones, and the forked sticks to the straight. Vegetation, short sticks, and straight sticks were more frequently collected in the second half of nest building than they were in the first half. In contrast to the males, sticks collected by females seemed a random sample of those available, and most were collected after the egg had already been laid. Comparison of a dissected nest with the notes taken on it during the experiment showed no significant differences. Pairs showed a wide range of variation in such things as length of time spent building and time of day spent building. Comparison of experienced and inexperienced builders proved inconclusive.

Chapter Six

Nest Defense

The territorial behavior of most colonial birds differs from that of their noncolonial relatives. Colonial species tend to have a restricted form of territoriality: the parents defend a small area around the nest (Welty 1975). Emlen (1954), for example, found that cliff swallows (*Petrochelidon pyrrhonota*) defended a territory the size of which was determined by the distance the swallow could reach with its bill while in the nest.

Wilson (1975) listed a variety of resources that territorial behavior could protect, for example, shelter, the food supply, nesting space, or space for sexual displays. Competition for these resources occurs mainly with conspecifics. Colonial birds seem to follow this pattern. Individuals defend their nests against their colony mates; predator defense is partially provided by the "safety in numbers" aspect of the whole colony. I investigated territorial behavior in the colonial Nicobars for 72 hours between 9 June and 21 August 1980 and 18 hours between 4 and 14 April 1981. This chapter will address two questions: Do these pigeons differentiate between intruders at their nests? and, What are they defending?

I considered an intruder to be any bird other than the owner approaching within an arbitrarily chosen three feet of a nest. Whenever possible, I noted the species of the intruder, the specific individual if it was a Nicobar, and the reaction of the bird on the nest. Foliage prevented me from seeing completely around every nest. Thus, these results are not based on every encounter between a nest owner and an intruder, but rather on those I could see. Because a threatening bird would draw my attention to a partially hidden intruder that I would not otherwise have seen, my results may include a higher proportion of threats than actually occurred. However, the potential for any such threats is assumed to have been distributed evenly over all classes of intruders.

Table 6-1 shows the identity of the intruders at each nest.

Responses to intrusions formed a continuum. I partitioned the 257 responses into two categories, threat and no threat.

1. Threat. In the most intense threat, the nest owner orients towards the intruder, stretches its neck, and jabs the open beak at the intruder. Sometimes the feathers are ruffled or the wings spread. Whitman (1919:142) described the Nicobar threat similarly, except he said that the feathers are always erect and that the threat is accompanied by a grunt. Less intense threats lacked some of the above elements.

2. No threat. The nest owner does not threaten the intruder, but instead ignores it (its behavior does not change and it shows no awareness of the intruder), looks at it (the owner rests its eyes briefly on the intruder), stares at it (the owner, its eyes wide open, looks fixedly at the intruder for many seconds), or becomes agitated (the owner, sometimes without even looking at the intruder, suddenly becomes more active -- it turns around, stands up and down, pokes at the nest, and/or looks around).

Table 6-1

Number of Intrusions at Nests of Five Pairs

	K22/K31 1980	55/41 1980	76/75 1980	86/35 1980	86/35 1981	K22/77 1981
	B, E, C	C	B, E, C	C	B	C
Breeding Females						
K24						
75					-	-
K31					-	-
41					2	1
77	2					
35						4
23						
Breeding Males						
76	4				-	-
K22			1			
68			1			
30	1		1		-	-
48	4					1
55						
86		1				11
K29					-	-
22						

(Table 6-1, cont'd.)

Non-Breeding Birds						
16 (adult)	11				-	-
24 (adult)					-	-
25		1				
80		1	1	1	-	-
5	2	2		1		
28	6			3	2	
1	-	-	-	-		
59	-	-	-	-		
64	-	-	-	-		
65	-	-	-	-		3
66	-	-	-	-		
Unidentified Nicobars	8	2	2	4	5	10
Touracos	28	21	2	1	1	
Other Species	26	37	16	16	2	7
TOTALS:	92	65	24	26	12	37

B = Building
E = Incubating
C = Rearing Chick

The two responses were not distributed randomly among the various intruders. The species of an intruder was a major influence on whether it was threatened or not (Table 6-2). Other Nicobars were the most likely to be threatened. When Nicobars were intruders, they were threatened 46% of 99 instances. Hartlaub's touracos (Tauraco hartlaubi) were threatened during only 19% of their 53 intrusions. Other intruding species received threats less than 4% of the time. These included shama thrushes (Copsychus malabaricus), golden breasted starlings (Cosmopsarus regius), superb starlings (Spreo superbus), violet-backed starlings (Cinnyricinclus leucogaster), ovenbirds (Seiurus aurocapillus), curly-crested helmet shrikes (Prionops plumata cristata), Sayaca tanagers (Thraupis sayaca), cattle egrets (Bubulcus ibis), and unidentified birds. The difference between the number of threats received by Nicobars and non-Nicobars is statistically significant (chi-squared = 47.72, 1 df, $P < .001$, phi = .43), as is the difference between the number of threats received by touracos and other non-Nicobar intruders (chi-squared = 9.88, 1 df, $P < .01$, phi = .25).

On six occasions, a bird between three and five feet away from a nest was threatened. Five of those threatened were Nicobars, and one was a touraco. These two species have already been shown to be the most likely to be threatened when within three feet.

The Nicobar intruders were not all threatened equally. Table 6-3 shows that 55% of the adult Nicobar intruders were threatened, compared to only 29% of immature Nicobars, a statistically significant difference (chi-squared = 5.0, 1 df, $P < .05$, phi = .25).

Table 6-2

Responses of Nesting Nicobars to Various Species of Intruders

	All Species	Nicobars	Touracos	All Others
No Threat	196	53	43	100
Threat	60	46	10	4
Totals	256	99	53	104

Table 6-3

Responses to Different Age Classes of Nicobar Intruders

	All Nicobars	Adult Nicobars	Immature Nicobars
No Threat	53	25	20
Threat	46	30	8
Totals:	99	55	28

The discrepancy between the sum of the number of adults and immatures, and the number of total intruders is due to those intruders not identified as to age class.

Table 6-4

Responses to Female and
Male Adult Nicobar Intruders

	All Nicobars	Female Nicobars	Male Nicobars
Threat	25	4	8
No Threat	30	5	17
Totals:	55	9	25

The discrepancy between the sum of the females and males, and the total number of adults is due to adults unidentified as to sex.

Table 6-5

Responses to Intruders During Different Stages of Nesting

		All Species	Nicobars	Non-Nicobars
Building Nest	No Threat	17	9	8
	Threat	17	16	1
	Total:	34	25	9
Incubating Egg	No Threat	39	3	36
	Threat	8	6	2
	Total:	47	9	38
Rearing Chick	No Threat	140	41	99
	Threat	35	24	11
	Total:	175	65	110

Sixty-eight percent of intrusions by males and 55% of intrusions by females resulted in threats (Table 6-4). This difference is not significant (chi-squared uncorrected for continuity = .42, 1 df, P < .70). Adult Nicobar 16, whose sex was unknown, accounted for 16 instances of intrusions by itself, with 45% resulting in a threat.

The distribution of defense varied through the breeding cycle. The nest owners were clearly more likely to threaten during the nestbuilding stage (50% of intruders were threatened) than during the incubation stage (17%) or the chick rearing stage (20%), as shown in Table 6-5. This difference between threat and no threat at different stages is highly statistically significant (chi-squared = 15.27, 2 df, P < .001, Cramer's V = .24).

The chick may also defend the nest. As it nears fledging, the parents begin to leave the chick alone at the nest, during which time it sometimes defends the nest. All three fledglings observed threatened at least one intruder. Whitman (1919) found that young ring doves (Streptopelia risoria) were aggressive near the nest, but "inoffensive" away from it. Kirkman (1937) twice observed young black-headed gull (Larus ridibundus) chicks repulse adults from their nests. Outside nests, however, chicks were pecked by adults, frequently to death. Similarly, when I observed fledged Nicobar young on the ground, I discovered their social interactions consisted mainly of being threatened, displaced, and chased.

The Nicobar pigeon could be defending the offspring or the nest itself. I concluded that the purpose of nest defense was to protect the nest. First, intruders were most likely to be threatened when the pair upon whose nest they intruded were still building, not when they had an egg or chick. Defense of the chick is characterized by increasing defense through the nesting cycle (Gowaty 1981). At the time of highest defense, the Nicobar had nothing to defend but the nest.

Second, the threats were not directed indiscriminately, but rather were directed towards adult Nicobars. Since the Nicobars breed synchronously, a Nicobar approaching a nest under construction would be likely to need sticks for its own nest. Thus, the threats were directed predominantly at those intruders most likely to steal nest material.

Third, the parents leave the chick alone and unprotected, even though it is still smaller than an adult and cannot fly, and it defends the nest.

Fourth, neither the Nicobars' natural habitat nor the Birdhouse free-flight area contains predators other than humans. The Nicobars themselves do not eat chicks. Perhaps the egg and chick are in little need of protection, while the scarce sticks are.

Fifth, although only 11 attempts were made to steal sticks from the observed nests, sticks do seem to be in high demand. Breeding males raided sticks from deserted nests, and nests that were abandoned were quickly disassembled. Nonbreeding and young Nicobars also collected and carried sticks. Any interest an intruder exhibited at a nest seemed directed towards the sticks rather than the occupants. Welty (1975) noted that stealing of nest materials occurs in penguins, cormorants, pelicans, storks, and rooks, among others, and that colonial nesters have the best opportunities to steal.

The Nicobars may also be defending their nest sites. Decreasing defense through the nesting cycle is consistent with defense of site (Gowaty 1981). However, the free-flight area contained numerous nest sites. Pairs fairly often chose sites never before used by Nicobars, even though many previously used locations were unoccupied and remained so. Also, pairs occasionally "tried out" more than one site before choosing one at which to build. Given the scarcity of sticks and the relative abundance of nest sites, Nicobar nest defense probably functions to protect the sticks, not the sites.

Nicobar nest defense probably has little to do with protection of genetic investment. (For a discussion of the theory of mate guarding, see Birkhead [1979], Lumpkin [1981], and Birkhead [1981].) As in many other species, the Nicobar females have the opportunity to copulate with males other than the mate, perhaps producing an egg fathered by the paramour. The mate could expend much energy caring for an unrelated chick. This problem is most likely to befall the inferior males; females only gain from extra-pair copulations when the partners are superior to the mate (Burley 1977:24). In a related species, the ring dove (S. risoria), males apparently insure their paternity by maintaining close social and visual contact with the female until the egg is laid (Lumpkin, et al. 1982). Among magpies (Pica pica), the males also shadow the females' movements, and chase away intruders (Birkhead 1979; Bultron 1983). Despite the superficial resemblance, Nicobar nest defense does not function to protect paternal investment. Females remain alone at the nest during building while males collect sticks. They usually do not occupy the nest together for very long during building. Male nest owners were on the nest during only 32% of intrusions during the building stage, and were almost always alone. Males could not possibly be protecting their mates from insemination by chasing off intruders from the nest.

Not only did females face most intruders by themselves, but during building they threatened 60% of Nicobar intruders (of which at most 44% were female), compared to 67% (of which half at most were males) threatened by males. Since forced copulations were never observed, the females had no need to fight off undesired suitors. They could not be defending their males, either, while alone at the nest.

Because the timing of Nicobar ovulation is unknown, the occurrence of mate guarding at another time cannot be ruled out. The second phase of courtship, for example, is characterized by physical proximity, and may be an example of mate guarding. But no evidence at all supports the idea that nest defense functions to protect the mate and not the nest.

The high number of threats received by touracos compared to other species is puzzling. Some threats might be credited to misidentification, since the touraco is roughly similar in size and body color to the Nicobar. They may also have been threatened more often due to their sudden flights past the Nicobar nests, which sometimes seemed to startle the nest owners.

In contrast to the aggression at the nests, the Nicobars were peaceful and gregarious on the ground. As with many pigeon species (Goodwin 1970), the Nicobars frequently flew, foraged, and rested in groups. Personal space maintained between birds on the ground was less than the three feet or more defended at the nest, sometimes being only about a foot. A sitting Nicobar would sometimes threaten a passerby with a thrust of its beak. However, most of what little aggression occurred away from nests took place at food bowls, and much consisted of running chases or fights in which each would attempt to hit the other with a wing, much like the defensive threat display of other pigeons described by Goodwin (1970).

Summary

I observed 256 intrusions at the nests of five pairs of Nicobars. The nest owners' responses to these intruders varied. Nest owners more frequently threatened conspecifics than other species, and adult Nicobars more than immatures. Recently fledged chicks also defended the nest. Threats were more likely during nest construction than during incubation or chick rearing. The function of nest defense in this colony seems to be protection of sticks in the nest, rather than protection of the offspring, mate, or nest site.

Chapter Seven

Incubation and Chick Rearing

Not much happens during incubation. The nest continues to be defended, but at a lower rate (Table 6-5). The partners take turns at incubation. (The relative proportion of time spent on the nest by each sex is currently being investigated by Kathleen Brown.) Incubation lasts for about 28 days.

Desertion of eggs may occur for several reasons. An egg that does not hatch will be deserted a few days after the due date. If one parent dies, the other deserts. If an egg is destroyed, the parents abandon the nest. Sometimes the parents desert for no apparent reason. When eggs from these nests can be rescued, they may be successfully hand-reared or cross-fostered. The rate of desertion in this colony cannot be estimated due to incomplete records.

Discovery of fallen eggshell is the major clue that a chick has hatched. Unlike the ring dove (Miller & Miller 1958:5) and feral pigeon (Goodwin 1983), the Nicobar parents do not fly away from the nest to drop the eggshell. This may be due to the lack of predators in their native habitat. In absence of eggshell, one must watch for increased parental restlessness and possible feedings. Weighing only 1/2 to 3/4 ounce, the altricial chick must grow much bigger before being visible above the nest edge (Plate 7-1). The parents take turns sitting on the chick and feeding it by regurgitation, as do other species of pigeons (Goodwin 1970:45). As it grows older, the chick sometimes defends the nest (Chapter 6).

Plate 7-1

Two-day-old Nicobar Pigeon

Only two chicks fledged during my 1982 observations. Chick 89 hatched 8 April and 91 hatched 15 April. Employing Focal Animal Sampling (Altmann 1974:242), I watched these chicks alternately for 1/2-hour samples each. Between 7 May and 14 June 1982, I watched chick 89 for 69 samples and chick 91 for 66 samples. I watched 89 for an additional 4-1/2 hours the day it first came to the ground, and 3-1/2 hours on a day when 91 had been temporarily removed. Between 7 July and 14 July 1982, I watched each chick for six samples. I noted such things as when the chick was left alone and for how long, when it first left the nest, and with whom it interacted. This Chapter also includes results from *ad libitum* sampling.

89 was first left alone at the nest for a few seconds at age 24 days, and 91 for a few minutes at age 25 days. On succeeding days, the chicks were occasionally left alone again. After an additional week and a half, the chicks were frequently left alone. 89 climbed out of the nest for about one minute at age 29 days. 91 stepped out of its nest with one foot at age 30 days, and with both feet at age 32. Once the chick had left the nest for the first time, it climbed out frequently, staying within a couple of feet of the nest.

I first saw 89 fly at the age of 40 days, and 89 flew to the ground for the first time at age 41 days. I observed 91 half-flying, half-jumping from branch to branch at age 36 days. 91 came to the ground for the first time at age 43 days, for the second time at age 45, and for good at age 46 days. 89 remained on the ground after coming down at age 41 days.

The young Nicobar in the nest begs for food by poking its bill up at or into the parent's bill. In 1980 I saw recently fledged Nicobars beg from their parents on the ground by quivering or fluttering their wings and squeaking. (Post-fledging feral pigeons also beg food by squeaking; however, this call is used before fledging as well [Burley 1977:40].) Neither 89 nor 91 begged from or was fed by their parents on the ground. However, at age 40 days, I saw 89 fluttering its wings. I also saw 91 fluttering at ages 34 and 37 days, and fluttering and squeaking at age 45 days. Thus, they developed the same behavior as the earlier chicks, but did not use it outside the nest.

91's father, 48, died when 91 was 27 days old. Female 77 did not desert 91, but rather increased her own sitting to compensate for 48's absence. As can be seen above, 91 developed behaviorally at about the same speed as 89. 91's weight during its father's illness (317 g at 24 days) and after his death (367 g at 30 days) was lower than almost all chicks weighed at similar ages.

From watching 89 only, one would think coming to the ground is a traumatic event for the Nicobar chick. From watching 91, however, one would think it instead an exciting adventure. On 89's first day down, it crashed into glass or walls at least six times. 89 was chased and attacked by several species of birds, and was the loser in one of the most violent fights I have ever seen between two Nicobars.

91, in contrast, seemed to adjust much more smoothly. It was never seen to fly into walls or glass. It came down the first time only briefly (Kathleen Brown, personal communication). The second day, it was on the ground 4 minutes, returning to the nest when chased by another Nicobar. I did not observe 91 on the day it finally came to the ground for good; however, 91 apparently got along fine with the other birds (Kathleen Brown, personal communication). On the first day 89 came down, the other Nicobars seemed curious: they stood near 89, and flew up and looked at it. I do not know if they were equally inquisitive about 91.

From the first day 91 was permanently on the ground, 89 and 91 spent much time together (Kathleen Brown, personal communication). They did not interact much. Rather, they foraged together, followed each other, and sat, stood, and bathed near each other. Out of the 44 Focal Animal Samples taken after both were permanently on the ground, 89 and 91 associated in the above ways in 25 of them. In addition, I frequently saw them together when I was not taking samples.

Neither chick interacted with its parents after leaving the nest. Although adults threatened 89 and 91 occasionally, 53 chased and threatened 89 and 91 frequently. 53 was the youngest of all the other Nicobars, and was just beginning to acquire adult plumage.

After both chicks were on the ground, the adults went through courtship and breeding again. Neither chick was involved in any courtship. 89 spent a few hours sitting on wire nest platforms, and once collected and carried a forked stick. This seems more "play" than real breeding behavior, and possibly was unrelated to the adult activities. Two other behaviors of these chicks might also have been play: suddenly running for no apparent reason, and running while flapping the wings.

These two examples of Nicobar chicks becoming independent of their nests and parents cannot be considered typical. First, since only two Nicobars fledged during the study period, I have only a small sample without any way of knowing the variation possible among chicks. Second, abnormal incidents befell each chick. 91's father died when it was only 27 days old. 89's parents laid a new egg in the same nest approximately 19 May 1982. Either might have disturbed the chicks' behavior, and might account for why neither 89 nor 91 was fed on the ground: 77 may have been too busy catching up on eating after spending double time at her nest with 91, and 22 and 23 may have lost interest in 89 after they had a new egg.

Although among pigeons both parents care for the chick, 48's death demonstrates that when one parent dies late in the nesting period, the other parent can finish raising the chick by itself. How old the chick must be for the remaining parent not to desert is unknown. However, on every occasion in this colony when one parent has died or deserted during incubation, the other parent eventually deserted the egg.

As Bell (1981) noted with hand-raised Nicobar chicks, behavioral development seems correlated with age, not physical size. Thus 91 nearly kept pace with 89 (see Figure 7-2), despite being small for its age.

Summary

I observed two chicks. They were first left alone on their nests when 3-1/2 weeks old, and left it themselves when a little more than 4 weeks old. They came to the ground at age 6 weeks. Unlike other chicks, these two were not fed by parents on the ground. The two chicks found each other after both were on the ground, and spent much time together.

FIGURE 7-2

COMPARISON OF BEHAVIORAL DEVELOPMENT OF FLEDGLINGS

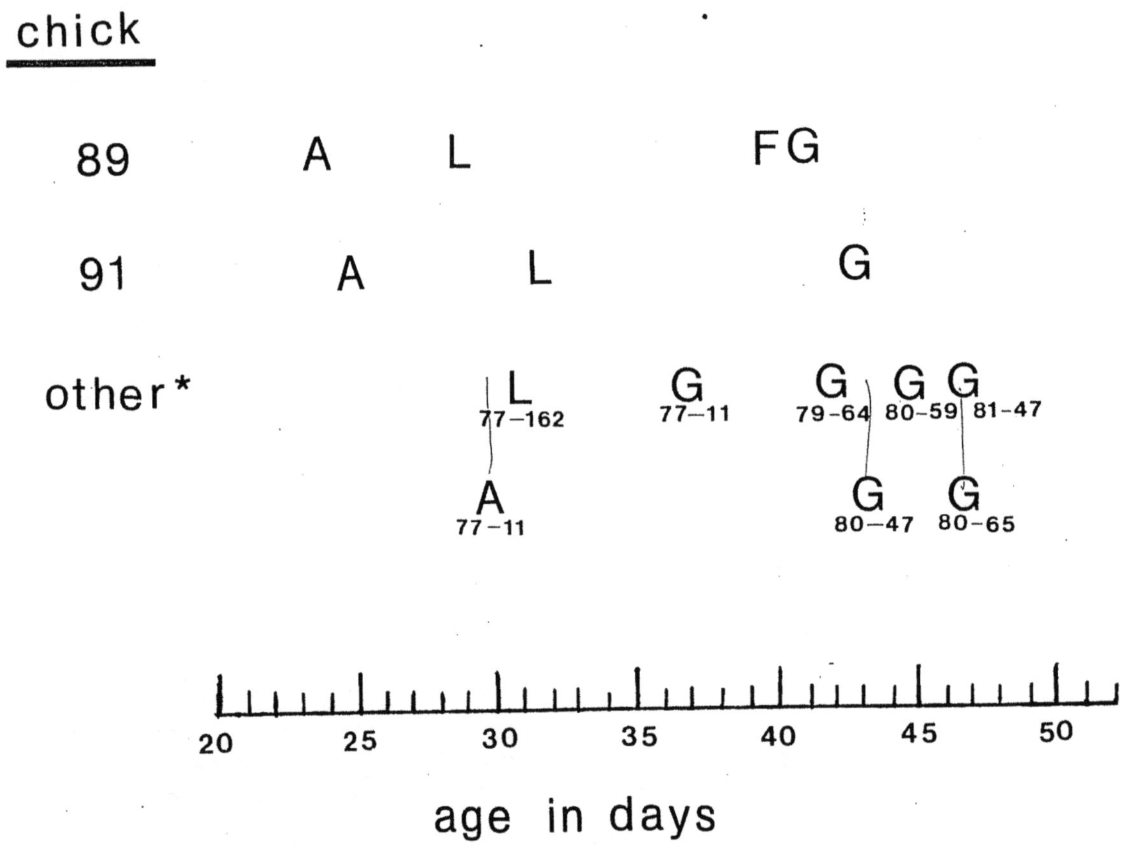

A = left alone on nest F = flew

L = left nest G = came to ground

*approximate ages, LCB under letter

Chapter Eight

Conclusions and Recommendations

The Nicobar pigeon is a beautiful bird; aesthetics seem reason enough to save it from extinction. But it is valuable for scientific reasons as well. The unique stomach, with its tough grinding surfaces and ingested pebble, has not been fully explained. Because of the Nicobar habit of traveling across the ocean, genetic studies of isolated island populations would be a valuable addition to the literatures of genetic drift and of island biogeography. The puzzling phenotypic homogeneity over the vast range, which I suggested (Chapter 4) might result from assortative mating by age, should be investigated. Because few other columbids nest colonially, the Nicobar pigeon is useful for comparative studies of nesting.

The results of this study provide the basis for several recommendations both for those attempting to breed Nicobars and for those interested in preventing its disappearance.

1. Adequate provisioning of sticks is a must. Nestbuilding at Lincoln Park Zoo lags when few sticks are available, and picks up when more are added (Kathleen Brown, personal communication). Those with aviaries containing other birds should be especially careful to provide enough sticks to compensate for those these species use. Plenty of forked sticks over a foot long should be provided, since these are the most popular. Other sticks and pieces of vegetation are needed as well. The Nicobars prefer to line the nest with different materials than formed the base; also, individuals can differ noticeably in preferences.

2. The Lincoln Park Zoo custom of lashing wire mesh platforms to forks of trees would seem of value to other breeders. I observed the Nicobars' difficulty getting a base started when not using a platform; I never observed any hindrance caused by a platform.

3. Although Bell (1981:218) suggested that increased density led to increased productivity, I could find no effect. Using Spearman's Rank Correlation Coefficient, I correlated various measures of density (number of monogamous pairs that could be formed of breeding birds, number of monogamous pairs that could be formed of birds one year old or older, total number of breeding birds, total number of birds one year old or older) with measures of productivity (number of eggs, number of successful eggs). The highest correlation, R_S = .67, was found between the number of breeding adults and number of eggs produced. However, even if the correlation were significant (it is not; Z = 1.64, NS), it would not be useful to the breeder. Without knowing how many of the young birds will join the ranks of breeding birds, one cannot predict whether the numbers of breeding birds will be high or low. Also, the number of chicks fledged seems a much better measure of productivity than number of eggs. R_S for number of birds one year old or older and number of successful eggs was only .06, again not significantly greater than zero. The crowding experienced by the Lincoln Park Zoo Nicobars may adversely affect breeding success in that increasing the number of birds does not significantly increase the number of young produced. This conclusion is surprising, given that the Nicobars are colonial.

Whether a given number of Nicobars will produce more young if divided into several small groups, or if concentrated into a few large groups cannot be answered from these data either. The Spearmans' Rank Correlation Coefficient of the number of birds one year or older with the average number of successful eggs per bird was -.47 (Z = -1.15, NS). Correlations of other measures of density and productivity also yielded correlations not significantly different from zero.

4. Using the data in Chapter 4, one can aid the birds in obtaining more highly preferred mates and possibly increase breeding success. Domestic pigeons allowed to choose their own mates laid earlier, laid more eggs, and had more fertile eggs than pigeons randomly paired by experimentors (Klint & Enquist 1981). Ring dove (S. risoria) pairs breeding together for the third time hatched more eggs than newly formed pairs of equal experience (Erickson & Morris 1972). Since the Nicobars prefer to mate with those the same age and with previous mates, efforts should be made to keep age cohorts together, either in their natal colony or when divided up among other colonies. Cohorts should only be broken up when the number of each sex are grossly asymmetrical. Established pairs should not be split either.

5. Even though the Nicobars breed all year, disturbances can be scheduled to minimize egg loss. I found the chance of breeding success varied throughout the year. During 1976 to 1982, 48 matings occurred with eggs due to hatch from April to July, while only 32 matings occurred the rest of the year (Figure 8-1). 27 peak-season eggs were successful, compared to only 11 off-season eggs. These

differences were not quite significant (chi-squared = 3.68, 1 df, .05 < P < .10, phi = .21). Of the unsuccessful eggs, a significantly greater number of off-season eggs were infertile (chi-squared corrected for continuity = 5.13, P < .05, phi = .34). Thus, if birds need to be weighed, medicated, moved, or otherwise bothered, doing it between August and March should minimize loss of potentially successful eggs.

6. Although male and female fledging weights overlap, that the average weights are significantly different (Chapter 2) means that some birds can be sexed at fledging. Birds whose weight at fledging falls above the male regression line (Y = 9.49X + 182.68, where Y is weight in grams, and X is age in days) are more likely to be males; those below the female regression line (Y = 6.74X + 204.74) are more likely to be females. These predictions do not apply to hand-raised birds, whose weights tend to be lower than parent-raised birds (Bell 1981).

FIGURE 8-1

EXPECTED HATCHDATES OF EGGS LAID 1976 – 1982

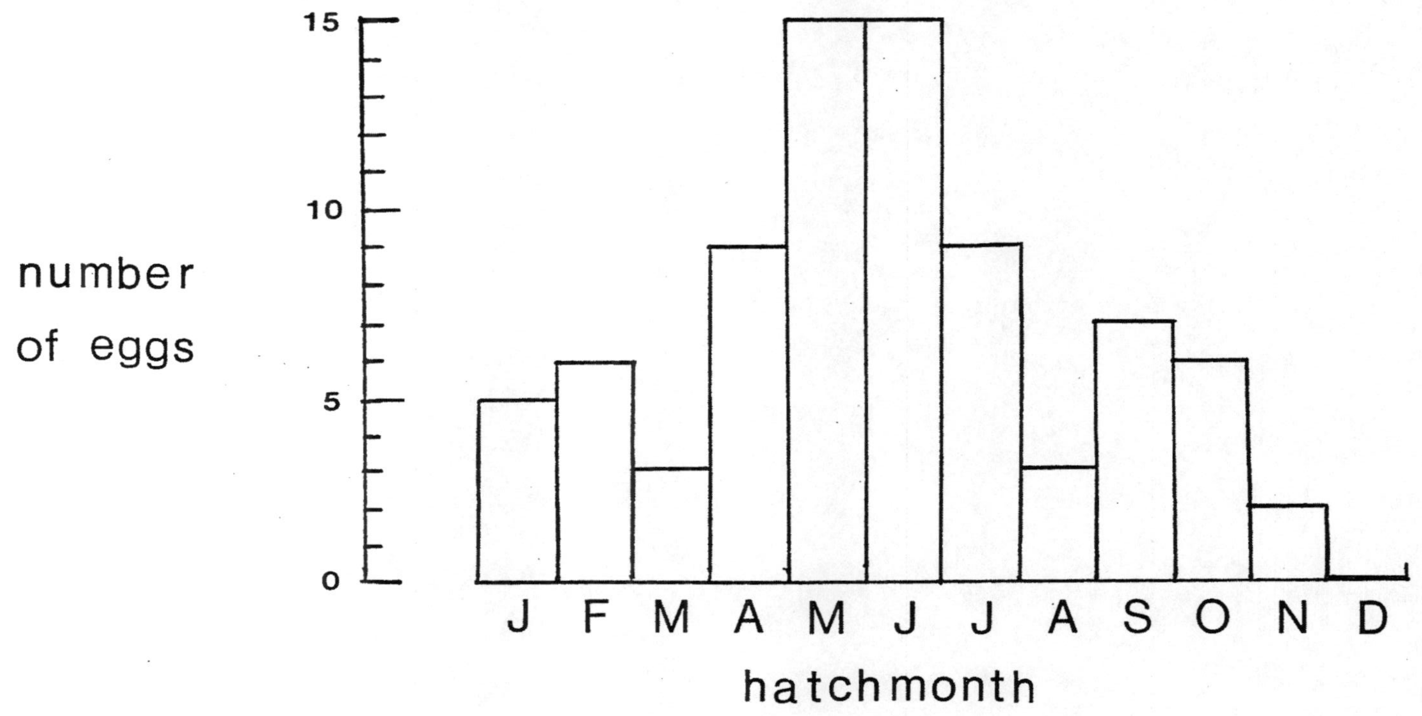

References

Ali, S. & Ripley, S. D. 1981. Handbook of Birds of India and Pakistan. Delhi: Oxford University Press.

Allen, E. F. 1951. The Nicobar Pigeon. Malay. Nat. J., 5, 151-155.

Altmann, J. 1974. Observational study of behaviour: sampling methods. Behaviour, **49**, 227-267.

Baker, E. C. S. 1913. Indian Pigeons and Doves. London: Witherby and Co.

Baker, E. C. S. 1928. The Fauna of British India, 2nd edition. Volume V, Birds. London: Taylor and Francis.

Baker, M. C. 1982. Vocal dialect recognition and population genetic consequences. Amer. Zool., **22**, 561-569.

Baker, M. C. & Mewaldt, L. R. 1978. Song dialects as barriers to dispersal in white-crowned sparrows, *Zonotrichia leucophrys nuttalli*. Evolution, **32**, 712-722.

Barlow, G. W., Rogers, W., & Cappeto, R. 1977. Incompatibility and assortative mating in the Midas cichlid. Behav. Ecol. Sociobiol., **2**, 49-59.

Bartels, E. 1964. Vragen rond het merkwaardige leven van de manenduif Caloenas nicobarica (L.). Ardea, **52**, 93-98.

Bateson, P. 1983. Optimal outbreeding. In: Mate Choice (Ed. by P. Bateson), pp. 257-277. Cambridge: Cambridge University Press.

Bell, D. J. 1983. Mate choice in the European rabbit. In: Mate Choice (Ed. by P. Bateson), pp. 211-223. Cambridge: Cambridge University Press.

Bell, K. J. 1981. Breeding and hand-rearing the Nicobar Pigeon *Caloenas nicobarica* at the Lincoln Park Zoological Gardens. Int. Zoo Yb., **21**, 217-219.

Berven, K. A. 1981. Mate choice in the wood frog, *Rana sylvatica*. Evolution, **35**, 707-722.

Birkhead, T. R. 1979. Mate guarding in the magpie Pica pica. Anim. Behav., **27**, 866-874.

Birkhead, T. R. 1981. Mate guarding in birds: conflicting interests of males and females. Anim. Behav., **29**, 304-305.

Bischof, N. 1972. The biological foundations of the incest taboo. Soc. Sci. Inform., **11**, 7-36.

Boesiger, B. 1982. Effets de la consanguinité sur des caracterès morphologiques quantitatifs chez Culex pipiens autogenicus Roub. Genetica, **59**, 99-111.

Brockelman, W. & Nadee, N. 1977. Preliminary survey and biogeographical analysis of the birds of the Surin Islands, Thailand. Nat. Hist. Bull. Siam. Soc., **26**, 211-226.

Brown, K. 1983. Captive management of the Nicobar pigeon at the Lincoln Park Zoo. Paper presented at American Federation of Aviculture meeting, Chicago, Illinois.

Bultron, D. 1982. Extra-pair courtship in blackbilled magpies. Anim. Behav., **31**, 211-220.

Burley, N. 1977. Mate choice and sexual selection in the pigeon, Columba livia. Ph.D. Thesis, The University of Texas at Austin.

Burley, N. 1980. Clutch overlap and clutch size: alternate and complementary reproductive tactics. Amer. Nat., **115**, 223-246.

Burley, N. 1981. Mate choice by multiple criteria in a monogamous species. Am. Nat., 117, 515-528.

Butler, A. L. 1899. The birds of the Andaman and Nicobar Islands. J. Bombay Nat. Hist. Soc., **12**, 684-696.

Cooke, F., Finney, G. H., & Rockwell, R. F. 1976. Assortative mating in lesser snow geese (Anser caerulescens). Behav. Genetics, **6**, 127-140.

Cothran, E., Chesser, R., Smith, M., & Johns, P. 1983. Influences of genetic variability and maternal factors on fetal growth in white-tailed deer. Evolution, **37**, 282-291.

Coulson, J. C. 1966. The influence of the pair-bond and age on the breeding biology of the kittiwake gull Rissa tridactyla. J. Anim. Ecol., **35**, 269-279.

Damon, A. 1977. Human Biology and Ecology. New York: W. W. Norton and Company.

Delius, J. D. & Emmerton, J. 1979. Visual performance of pigeons. In: Neural Mechanisms of Behavior in the Pigeon (Ed. by A. M. Granda and J. H. Maxwell), pp. 51-70. New York: Plenum Press.

Denniston, C. 1978. Small population size and genetic diversity: implications for endangered species. In: Endangered Birds: Management Techniques for Preserving Threatened Species (Ed. by S. A. Temple), pp. 281-289. Madison: University of Wisconsin Press.

Downhower, J. F. & Brown, L. 1981. The timing of reproduction and its behavioral consequences for mottled sculpins, Cottus bairdi. In: Natural Selection and Social Behavior (Ed. by R. D. Alexander & D. W. Tinkle), pp. 78-95. New York: Chiron Press.

Emlen, J. T. 1954. Territory, nestbuilding and pair formation in the cliff swallow. Auk, **71**, 16-35.

Erickson, C. J. & Morris, R. L. 1972. Effects of mate familiarity on the courtship and reproductive success of the ring dove (Streptopelia risoria). Anim. Behav., **20**, 341-344.

Fabricius, E. & Jansson, A. 1963. Laboratory observations on the reproductive behavior of the pigeon (Columba livia) during the pre-incubation phase of the breeding cycle. Anim. Behav., **11**, 534-547.

Flower, W. H. 1860. On the structure of the gizzard of the Nicobar Pigeon, and other granivorous birds. Proc. Zool. Soc. Lond., **28**, 330-334.

Freund, J. E. & Walpole, R. E. 1980. Mathematical Statistics, 3rd edition. Englewood Cliffs, NJ: Prentice-Hall, Inc.

Frith, H. J. 1977. Some display postures of Australian pigeons. Ibis, **119**, 167-182.

Goforth, W. R. & Baskett, T. S. 1971. Social organization of penned mourning doves. Auk, **88**, 528-542.

Goodwin, D. 1970. Pigeons and Doves of the World. London: Trustees of the British Museum.

Goodwin, D. 1983. Behavior. In: Physiology and Behavior of the Pigeon (Ed. by M. Abs), pp 285-308. New York: Academic Press.

Gowaty, P. A. 1981. Aggression of breeding eastern bluebirds (*Sialia sialis*) towards their mates and models of intra- and interspecific intruders. Anim. Behav., **29**, 1013-1027.

Grau, H. J. 1982. Kin recognition in white-footed deermice (*Peromyscus leucopus*). Anim. Behav., **30**, 497-505.

Greenwood, P. J., Harvey, P. H., & Perrins, C. M. 1978. Inbreeding and dispersal in the great tit. Nature, Lond., **271**, 52-54.

Halliday, T. R. 1983. The study of mate choice. In: Mate Choice (Ed. by P. Bateson), pp. 3-32. Cambridge: Cambridge University Press.

Heinroth, O. 1902. Ornithologische ergebnisse der "I. Deutschen Südsee Expedition von Br. Mencke." J. Orn., Lpz., **4**, 390-457.

Hieber, C. S., and Cohen, J. 1983. Sexual selection in the lovebug, *Plecia nearctica*: the role of male choice. Evolution, **37**, 987-992.

Hoogerwerf, A. 1967. Verspreiding en voedsel van *Caloenas nicobarica* en *Ducula bicolor*. Ardea, **55**, 249-259.

Hoogland, J. L. 1982. Prairie dogs avoid extreme inbreeding. Science, N.Y., **215**, 1639-1641.

Hopkinson, E. 1926. Records of Birds Bred in Captivity. London: H. F. and G. Witherby.

Hume, A. O. 1874. Contributions to the ornithology of India: the islands of the Bay of Bengal. Stray Feathers, **2**, 29-324.

Hume, A. O. 1890. The Nests and Eggs of Indian Birds, 2nd edition, volume II (Ed. by E. W. Oates). London: R. H. Porter.

Johnson, L. K. 1982. Sexual selection in a brentid weevil. Evolution, **36**, 251-262.

King, W. B., compiler, 1981. Endangered Birds of the World: the ICBP Bird Red Data Book. Washington, DC: Smithsonian Institution.

Kirkman, F. B. 1937. Bird Behavior: A Contribution Based Chiefly on a Study of the Black-headed Gull. London: T. Nelson and Sons, Ltd.

Klint, T. & Enquist, M. 1981. Pair formation and reproductive output in domestic pigeons. Behav. Processes, **6**, 57-62.

Kreithen, M. L. 1979. The sensory world of the homing pigeon. In: Neural Mechanisms of Behavior in the Pigeon (Ed. by A. M. Granda and J. H. Maxwell), pp. 21-33. New York: Plenum Press.

Lehrman, D. S. 1961. Hormonal regulation of parental behavior in birds and infrahuman mammals. In: Sex and Internal Secretions, Vol. 2 (Ed. by W. C. Young), pp. 1268-1382. Baltimore: Williams and Williams Co.

Lehrman, D. S. 1964. Control of behavior cycles in reproduction. In: Social Behavior and Organization among Vertebrates (Ed. by W. Etkin), pp. 143-166. Chicago: University of Chicago Press.

Lenington, S. 1980. Female choice and polygyny in redwinged blackbirds. Anim. Behav., **28**, 347-361.

Lenington, S. 1983. Social preferences for partners carrying "good genes" in wild house mice. Anim. Behav., **31**, 325-333.

Levitan, M. & Montagu, A. 1971. Textbook of Human Genetics. New York: Oxford University Press.

Lewin, R. 1983. No dinosuars this time. Science, N. Y., **221**, 1168-1169.

Linné, C. von. 1758. Systema Naturae, 10th edition. London: Trustees of the British Museum. 1939 facsimile.

Lorenz, K. Z. 1981. The Foundations of Ethology. New York: Springer-Verlag.

Lumpkin, S. 1981. Avoidance of cuckoldry in birds: the role of the female. Anim. Behav., **29**, 303-304.

Lumpkin, S., Kessel, K., Zenone, P., & Erickson, C. J. 1982. Proximity between the sexes in ring doves: social bonds or surveillance? Anim. Behav., **30**, 506-513.

Mayr, E. 1945. Birds of the Southwest Pacific. New York: MacMillan and Company.

Mayr, E. 1946. History of the North American bird fauna. Wilson Bull., **58**, 3-41.

Mayr, E. 1972. Sexual selection and natural selection. In: Sexual selection and the Descent of Man 1871-1971 (Ed. by B. Campbell), pp. 87-104. Chicago: Aldine Publishing Company.

McClintock, M. K., Anisko, J. J., & Adler, N. 1982. Group mating among Norway rats II. The social dynamics of copulation: competition, cooperation, and mate choice. Anim. Behav., **30**, 410-425.

Miller, W. J. & Miller, L. S. 1958. Synopsis of behavior traits of the ring dove. Anim. Behav., **6**, 3-8.

Nicolai, J. 1962. Über Regen-, Sonnen- und Staubadden bei Tauben (Columbidae). J. Orn., Lpz., **103**, 125-139.

Noordwijk, A. J. van & Scharloo, W. 1981. Inbreeding in an island population of the great tit. Evolution, **35**, 674-688.

Packer, C. 1979. Inter-troop transfer and inbreeding avoidance in Papio anubis. Anim. Behav., **27**, 1-36.

Parker, S. 1976. The precultural basis of the incest taboo: toward a biosocial theory. Amer. Anthrop., **78**, 285-305.

Partridge, L. 1983. Non-random mating and offspring fitness. In: Mate Choice (Ed. by P. Bateson), pp. 227-255. Cambridge: Cambridge University Press.

Petrie, M. 1983a. Female moorhens compete for small fat males. Science, **220**, 413-415.

Petrie, M. 1983b. Mate choice in role-reversed species. In: Mate Choice (Ed. by P. Bateson), pp. 167-179. Cambridge: Cambridge University Press.

Pratt, H. D., Engbring, J., Bruner, P. L., & Berrett, D. G. 1980. Notes on the taxonomy, natural history, and status of the resident birds of Palau. Condor, **82**, 117-131.

Pusey, A. E. 1980. Inbreeding avoidance in chimpanzees. Anim. Behav., **28**, 543-552.

Ralls, K., Brugger, K., & Ballou, J. 1979. Inbreeding and juvenile mortality in small populations of ungulates. Science, N. Y., **206**, 1101-1103.

Rand, A. L. & Gilliard, E. T. 1967. Handbook of New Guinea Birds. London: Weidenfeld and Nicolson.

Reichenbach, H. G. L. 1847-1862. Die vollständigste Naturgeschichte der Tauben und Taubenartigen Vögel. Dresden and Leipzig. (No publisher listed.)

Rowland, W. J. 1982. Mate choice by male sticklebacks, Gasterosteus aculeatus. Anim. Behav., **30**, 1093-1098.

Roychoudhury, A. K. & Sankhala, K. S. 1979. Inbreeding in white tigers. Proc. Indian. Acad. Sci., **88-B**, 311-323.

Rutowski, R. L. 1982. Epigamic selection by males as evidenced by courtship partner preferences in the checkered white butterfly (Pieris protodice). Anim. Behav., **30**, 108-112.

Sade, D. S., Chepko-Sade, D., Schneider, J., Roberts, S., and others. In preparation. Basic Demographic Observations on Free-ranging Rhesus Monkeys.

Salomonsen, F. 1975. The main problems concerning avian evolution on islands. Proc. 16th Intern. Ornith. Congr., 585-602.

Salvadori, T. 1893. Catalog of Birds in the British Museum, volume **21**. London: British Museum.

Sassaman, C. 1978. Dynamics of a lactate dehydrogenase polymorphism in the wood louse Porcellio scaber Latr.: evidence for partial assortative mating and heterosis in natural populations. Genetics, **88**, 591-609.

Selander, R. K. 1972. Sexual selection and dimorphism in birds. In: Sexual Selection and the Descent of Man 1871-1971 (Ed. by B. Campbell), pp. 180-230. Chicago: Aldine Publishing Company.

Shields, W. M. 1982. Philopatry, Inbreeding, and the Evolution of Sex. Albany: State University of New York Press.

Snedecor, G. W. & Cochran, W. G. 1968. Statistical Methods, 6th Edition. Ames, Iowa: Iowa State University Press.

Stills, S. 1970. Love the one you're with. Steven Stills. Atlantic Records, At 7202.

Temple, S. A. 1981. Applied island biogeography and the conservation of endangered island birds in the Indian Ocean. Biol. Conserv., **20**, 147-161.

Trivers, R. L. 1972. Parental investment and sexual selection. In: Sexual Selection and the Descent of Man 1871-1971 (Ed. by B. Campbell), pp. 136-179. Chicago: Aldine Publishing Company.

Tukey, J. W. 1977. Exploratory Data Analysis. Reading, MA: Addison-Wesley Publishing Co.

Verrell, P. A. 1982. Male newts prefer large females as mates. Anim. Behav., **30**, 1254-1255.

Wallace, A. R. 1865. On the pigeons of the Malay Archipelago. Ibis, **1**, 365-400.

Wallace, A. R. 1869. The Malay Archipelago. London: MacMillan and Company. 1962 reprint. The Malay Archipelago. New York: Dover Publications, Inc.

Wallace. B. 1968. Topics in Population Genetics. New York: W. W. Norton and Company, Inc.

Weber, P. G. & Weber, S. P. 1976. The effect of female color, size, dominance and early experience upon mate selection in male convict cichlids, Cichlasoma nigrofasciatum, Gunther (Pisces, Cichlidae). Behaviour, **56**, 116-135.

Welty, J. C. 1975. The Life of Birds, 2nd edition. Philadelphia: W. B. Saunders Co.

Whitman, C. O. 1919. Posthumous works. The Behavior of Pigeons, volume 3. Washington, DC: Carnegie Institute Publication.

Wilson, E. O. 1975. Sociobiology. Cambridge, MA: Belknap Press.

Wolfe, E. D. B. 1951. The Nicobar Pigeon. Malay. Nat. J., **5**, 204.

Wright, S. 1977. Evolution and the Genetics of Populations, Vol. 3. Chicago: University of Chicago Press.

Update, June 2012

Since this dissertation was finished in 1983, few scholarly reports have been published on the Nicobar pigeon, *Caloenas nicobarica*. Fewer yet have focused on reproductive behavior. Thus, despite the intervening 29 years, this update will be brief and focus primarily on the factors threatening the continued existence of the Nicobar pigeon in the wild and how this dissertation may help its survival.

Prospects for Survival

The Nicobar pigeon is currently listed as "near threatened" (IUCN 2011; UNEP-WCMC) rather than "endangered." However, the population is declining for several reasons, including habitat destruction and the introduction of predators onto their islands (BirdLife International 2012).

Recent research has discovered several relatives of the Nicobar pigeon, one a very close relative. All lived on islands and none survive today. All became extinct after human incursion on their previously isolated islands. The fate of these relatives presents a sobering picture of what lies in store for the Nicobar pigeon unless humans intervene to preserve their habitat.

Discovery of Related Species

As mentioned above, there are now four species known to be related to the Nicobar pigeon.

In 1989, two research groups announced discoveries of bones of a previously unknown, now-extinct species of Caloenas, *Caloenas canacorum* (Hume and Walters 2012). The remains were found on New Caledonia and on an island in the Kingdom of Tonga. This new species has been variously called the giant Nicobar pigeon, the greater maned pigeon, the Kanaka pigeon, and the great green pigeon. Nothing of its reproductive behavior is known. Based on its remains, it was a big version of the living Nicobar pigeon.

The existence of the "Liverpool pigeon" has long been known from two specimens that were probably collected in the Pacific. (One of these specimens has since disappeared.) This pigeon is now thought to be another species of Caloenas. Now labeled *Caloenas maculata*, it probably lived on small, predator-free islands. Its short wings suggest that it did not migrate from island to island as *Caloenas nicobarica* does. It was probably smaller than *Caloenas nicobarica*. The Liverpool pigeon was declared extinct in 2008 (IUCN 2011) after decades of exploration of the Pacific failed to find a wild population. Nothing is known of the reproductive behavior of the Liverpool pigeon.

In 2002, a genetic analysis suggested that the closest living relative of the dodo (*Raphus cucullatus*) and the solitaire (*Pezophaps solitaria*) is the Nicobar pigeon (Shapiro *et al.* 2002). The dodo and the solitaire, both now extinct, were flightless island dwellers whose gizzards contained stones. Similarly, the Nicobar pigeon lives on islands, has gizzard stones, and is mainly a ground-living bird. Although uncommon, gizzard stones evolved in several lines of vertebrates, sometimes as a digestion aid and sometimes as ballast. In the dodo, the solitaire, and the Nicobar pigeon, gizzard stones grind hard foods and likely were present in their common ancestor.

Dodos and solitaires have long been of popular and scientific interest for reasons that include:

- They are among the bird species that lost the ability to fly while living on remote islands without predators.
- Both exhibit island gigantism. The dodo stood more than three feet tall and weighed an estimated 50 pounds. The solitaire probably weighed nearly as much. (Gigantism is also something these birds share with the genus Caloenas: The extinct *Caloenas canacorum* was 25% larger than the extant *Caloenas nicobarica.*)
- Early explorers found the gizzard stones surprising and unusual.
- After the dodo went extinct about 1690 C.E. (Hume *et al.* 2004), Western Europeans realized for the first time that human activities could cause the extinction of an entire species. The solitaire followed the dodo into extinction within 50 years.
- The dodo was once considered a mythical bird like the roc and the phoenix, possibly because of its unusual features and ugly, malproportioned body.
- Lewis Carroll made the dodo famous and beloved by having one as a character in his children's book *Alice's Adventures in Wonderland.*

However, despite the interest in these birds, little is known about them because only a handful of specimens, travelers' accounts, and drawings exist. Now that their genetic relationship to the Nicobar pigeon has been uncovered, the Nicobar pigeon's anatomy, physiology, and behavior can be good sources of hypotheses in answering questions about the dodo and solitaire.

Shapiro et al.'s (2002) genetic analyses also revealed that the bird family most closely related to the Columbiformes (the taxonomic order of pigeons and doves) is the Laridae, which includes shorebirds such as gulls and skuas. Because the Nicobar pigeon diverged very early from the rest of the Columbiformes, comparisons with the Laridae may benefit scholars of both the Nicobar pigeon and of some shorebirds.

Island Biogeography of the Nicobar Pigeon

Resources are limited on small islands. As a result, some island-dwelling species have evolved energy-saving traits to improve their odds of survival. McNab (1994) named three strategies:

- reduction in energy use through such means as reduced inclination to fly, reduced ability to fly, a shift from endothermy to ectothermy, reduced body size, and channeling more energy into reproduction and less into metabolism
- reduction in population fluctuations
- increase in population size

A later biogeographical analysis by McNab (2000) may explain why *Caloenas nicobarica* has survived when larger relatives with presumably similar habits have become extinct. McNab compared metabolism and temperature regulation in 16 species of pigeons that live in the South Pacific. Ten species lived on a mainland, and six species (including the Nicobar pigeon) lived on small islands. McNab found that in general, pigeons that lived on small tropical islands had a lower basal rate (energy expenditure) corrected for body mass than pigeons living on a temperate mainland. Among pigeons living on islands, metabolic rate correlated with island size, with pigeons on smaller islands having lower metabolic rates. These results suggest that pigeons on islands have indeed evolved to reduce energy use to better weather their limited resources and greater susceptibility to ecosystem changes.

McNab also found that some small-island pigeons have smaller bodies than their relatives on continents, reducing the island birds' energy requirements. He was unable to determine whether the island pigeons had evolved a lower metabolic rate as well.

The three Nicobar pigeons included in McNab's (2000) analysis came from small islands. They showed a wider (and statistically significant) variation in basal rate among individuals than in any other species of pigeon. One Nicobar pigeon had a basal rate 80% that of the average of all pigeons. Another had a basal rate 92% of average, and the third had a basal rate 104% of average.

Based on the average basal rate of these three individuals, the Nicobar pigeon appears to have evolved to use less energy. The wide variation among individuals hints that the species may still be evolving toward lower energy use and/or that this evolutionary trend started recently.

Mitchell (1989) reported briefly on one breeding season on two islands in Irian Jaya (now called Western Papua) in Indonesia. Nicobar pigeons and pied imperial pigeons (*Ducula bicolor*) routinely breed on these islands from October to March. In 1985, the smaller island, Pulau Kumbur (2 hectares), hosted 150 pigeons, most of them Nicobar pigeons. The larger island, Pulau Nutabari (4 hectares), had perhaps 300 pigeons during this time.

Mitchell (1989) observed that the Nicobar pigeons nested in trees and that their nests varied widely in height: from about 6 foot off the ground to about 40 feet high on both islands. The islands varied in topography, vegetation, and habitat disturbance. The breeding colony on Pulau Kumbur is threatened by hunters, who shoot many pigeons while they are clustered in large groups in a small area.

Mitchell's observations confirm the observations reported in this dissertation and and add to them. Even in the wild on uninhabited islands, Nicobar pigeons nest densely in trees just as they do in the zoo. In both the wild and in the zoo, Nicobar pigeons are not choosy about the species of tree they nest in. Bell (1981) suggested that higher bird density in the zoo led to greater breeding success.

However, the winter breeding season observed by Mitchell differs in timing and length from that in the Lincoln Park Zoo Nicobar pigeons. The latter breed year round, although most eggs (60.0%, according to data presented in this dissertation) hatch in April, May, June, and July. Also different, for obvious reasons, is that the wild Nicobar pigeons built some nests as high as 40 feet. The ability of wild Nicobar pigeons to spread out vertically as well as horizontally decreases nest concentration, leaving open the question of the optimal density of Nicobar pigeons during breeding season.

Other researchers have studied the effect of human intervention on bird populations on formerly isolated islands.

Holmes and van Balen (1995) visited two neighboring islands off the coast of West Java. Pulau Tinjil measures 600 hectares, and Pulau Deli measures 950 hectares. When the authors visited the two islands in 1988, Nicobar pigeons were common on both. A breeding colony of long-tailed macaques (*Macaca fascicularis*) had been established the year before on Pulau Tinjil, and another colony was established on Pulau Deli in 1989. By 1992, a monkey census estimated that about 2,000 monkeys lived on Pulau Tinjil. A monkey census in 1995 on Pulau Deli estimated 8,000 monkey inhabitants. When Holmes and van Balen revisited Pulau Deli for three days in 1995, they spotted only three Nicobar pigeons. They concluded that the Nicobar pigeon population may have declined because of the introduction of monkeys.

Steadman (1997) collated information on the number of pigeon and dove species in Polynesia in prehistoric times and today. He discovered that the islands of Polynesia currently support fewer species of pigeons and doves than before human incursion (which started about 1500 B.C.E.). Some pigeon species, including *Caloenas canacorum*, have become extinct in Polynesia. Other pigeon species survive, but inhabit fewer islands than in the past.

This trend is continuing, and small, declining populations of some species are expected to disappear altogether from some islands. Steadman attributes the decline and extinction of doves and pigeons in Polynesia after human colonization to several factors:

- deforestation
- introduction of foreign pathogens
- human hunting of pigeons and doves for fat, feathers, and meat
- clearing of native plants to make way for fields of non-native crops
- predation by dogs, rats, and pigs (introduced by humans) on both pigeons and their food

The observations of Holmes and van Balen (1995) and Steadman's analysis (1997) suggest that pigeons and doves native to uninhabited islands do not adjust well to human incursion. The fates of *Caloenas canacorum* and *Caloenas maculata* suggest that the genus Caloenas is no exception. The Polynesian populations of pigeons and doves are still declining after as much as 3,500 years of co-existance with humans.

The long-term survival of Nicobar pigeons and other island-dwelling pigeons and doves will likely require strong conservation efforts starting now. As yet there is no evidence that they can adapt and continue to breed successfully when their nesting islands are no longer isolated.

Conservation efforts are particular important for the subspecies *Caloenas nicobarica pelewensis*, whose tiny range is restricted to rocky islands in the southern part of the Kingdom of Palau.

References

Bell, K.J. 1981. Breeding and hand-rearing the Nicobar pigeon *Caloenas nicobarica* at the Lincoln Park Zoological Gardens. *International Zoo Yearbook* **21**: 217–219.

BirdLife International 2012. Species factsheet: *Caloenas nicobarica.* http://www.birdlife.org/datazone/speciesfactsheet.php?id=2604. Viewed on 17 May 2012.

Holmes, D., and S. (Bas) van Balen 1995. The birds of Tinjil and Deli Islands, West Java. *Kukila* **8**: 117–126.

Hume, J.P., et al. 2004. Palaeobiology: Dutch diaries and the demise of the dodo. *Nature* **429**: 622.

Hume, J.P, and Michael Walters 2012. *Extinct birds.* London: T. and A.D. Poyser.

IUCN 2011. *IUCN Red List of Threatened Species.* Version 2011.2. http://www.iucnredlist.org. Viewed on 17 May 2012.

McNab, B.K. 1994. Resource use and the survival of land and freshwater vertebrates on oceanic islands. *The American Naturalist* **144**: 643–660.

McNab, B.K. 2000. The influence of body mass, climate, and distribution on the energetics of South Pacific pigeons. *Comparative Biochemistry and Physiology*, **127**: 309–329.

Mitchell, A. 1989. Nicobar pigeon nesting colonies in Irian Jaya. *Kukila* **4**(1–2): 51–52.

Shapiro, B. et. al. 2002. Flight of the dodo. *Science* **295**: 1683. Supplementary material not in the hardcopy magazine is available online at http://www.sciencemag.org/content/suppl/2002/02/28/295.5560.1683.DC1.

Steadman, D.W. 1997. The historic biogeography and community ecology of Polynesian pigeons and doves. *Journal of Biogeography* **24**: 737–753.

UNEP-WCMC 2012. *UNEP-WCMC Species Database: CITES-Listed Species.* Viewed on 17 May 2012.

Appendix I

Ethogram of the Nicobar Pigeon

Following are the categories of behaviors I found useful when observing the Nicobar pigeons, accompanied by descriptions. These behaviors are not all at the same level of organization. For example, Tremble-shoving and Bowing are probably fixed action patterns, while Bathing and Chasing are more complex behaviors comprising simpler units in varying orders and frequencies.

<u>Adjusting Nest</u>: A Nicobar in its nest bites or tremble-shoves pieces of the nest that project from the sides or top. This can occur during any stage of nesting.

<u>Allopreening</u>: Two Nicobars bite at and comb through each other's feathers. The most frequently groomed parts are the head and neck ruff. The partner may draw the entire length of a long hackle through its beak. Usually the Nicobars are of opposite sexes. The female tends to initiate preening; the partner usually then responds in kind. (cf. Goodwin 1970:35-36; Miller & Miller 1958:5)

<u>Bathing</u>: The Nicobar lies in the rain or in the sun, stretching its wings and lifting them to expose the side and underwing feathers. The feathers are fluffed out. The bird changes position frequently. (cf. Goodwin 1970:22-23; Miller & Miller 1958:7)

<u>Bowing</u>: The Nicobar quickly rotates the head downward so that the beak touches or almost touches the chest and is parallel to the ground, then raises it again. Bows are very stereotyped, occur in trees, and last roughly 0.5 to 1.0 second. Bows are sometimes isolated, but more usually occur in sequences. Sometimes the tail is slightly spread, and sometimes the wings are lowered. Males do most of the bowing. (cf. Goodwin 1970:28-30; Frith 1977:173-179; Miller & Miller 1958:5)

<u>Chasing</u>: One Nicobar moves rapidly in pursuit of another. A few brief chases were aerial. Most chases involved one Nicobar running after another on the ground. Chases lasted as long as five minutes. A chase ends when the chasee flies away or the chaser stops running.

<u>Chest Dipping</u>: A Nicobar stands or squats with its chest, and consequently its head, lower than its tail. This occurs during interactions between two birds. A female on a nest with her mate usually keeps her head well below his. In a fight, one Nicobar may stand with chest dipped. This may signal submission.

<u>Courtship Begging</u>: The Nicobar puts its chest low and lifts its beak up to its partner's beak. Sometimes it forces its beak into the other's and tries to pry

its mouth open. The other Nicobar may pull its head back out of reach, voluntarily open its beak and regurgitate crop milk, or may regurgitate when the other forces its beak down its throat. Almost always the female is the begger, and the male then is the feeder. Mounting by the feeder frequently follows begging. (cf. Goodwin 1970:36; Miller & Miller 1958:6)

Feeding Squab: The Nicobar parent opens its bill and allows the squab to insert its bill. The parent chokes as it regurgitates crop milk or food. Both males and females feed squabs.

Following: One Nicobar walks behind another, usually of the opposite sex. Either sex can follow. (cf. Burley 1977:49)

Foraging: The Nicobar walks slowly with no apparent destination, looking at the ground and poking, picking up, dropping, or eating things. (cf. Goodwin 1970:19)

Half Stretching: The Nicobar lifts both wings up above the body at once. The wings are folded so that the edges, at shoulder height, are lower than the rest of the wings. (cf. Goodwin 1970:23)

Head Scratching: The Nicobar lifts its foot in front of the wing and scratches its head on the same side. Usually the eyes are closed. (cf. Goodwin 1970:23; Miller & Miller 1958:8)

Head Throwing: The Nicobar lowers its chest, raises its hind quarters, and throws the head upwards, adding an extra jerk at its highest point, then lowers it again. Head Throwing is quite variable, and sometimes grades into Nodding. Only males were seen to throw their heads.

Hopping: The Nicobar jumps up into the air. This takes place during a fight, and the bird lands closer to the opponent, occasionally landing on the opponent itself. (cf. "standing on opponent" Miller & Miller 1958:6)

Incubating: A Nicobar, rarely moving, sits low in the nest on an egg. Both males and females incubate.

Lying Down: The Nicobar rests on the substrate, supported either totally by its side, or partially by its stomach as well. Birds lie down either on the ground or in trees, where they hold onto a branch with their toes. Lying down occurs during bathing. Fledglings also sometimes lie down when not bathing. (cf. "relaxed resting" Miller & Miller 1958: 7).

Lowered-Wing Walking: The Nicobar walks stiffly with its primaries spread apart and wings lowered so that they hit or almost hit the ground or branch with each step. The white tail may be lowered as well. Lowered-Wing Walking was usually done by males. (cf. "parading" Goodwin 1970:32)

Mounting: A Nicobar hops onto another's (usually the mate's) back and stands there. Mounting is sometimes followed by copulation. Usually males mount.

Nodding: The Nicobar, in a tree, its head somewhat extended, moves it up and down, usually without any apparent rotation. Thus, the beak remains parallel to the ground. Usually the chest is low and the tail up. Nodding grades into Head Throwing. Only males were observed nodding. (cf. Goodwin 1970:32-34)

Preening: The Nicobar cleans it feathers by poking them, digging among them, pulling them, or running them through the beak. (cf. Goodwin 1970:23)

Ritual Looking: The Nicobar thrusts its head out far, pauses briefly as if looking, retracts the head, then repeats this in different directions. The head is not always thrust out the same distance nor for the same length of time. Both males and females performed this behavior, both on the ground and up in trees. (cf. "peering" Frith 1977:172; "rapid peering" Miller & Miller 1958:7)

Running: The Nicobar locomotes by moving first one leg forward, then the other, quickly.

Running At: A Nicobar on the ground runs quickly at another Nicobar. The latter's response is usually to run at least a few steps away.

Shaking Head and/or Ruff: The Nicobar moves its head rapidly laterally, so that the hackles may fly in the air. Physical irritation of the head may cause Shaking. (cf. Goodwin 1970:38-39)

Shifting on Nest: A nesting Nicobar turns on the nest, leans back and forth, steps on the nest, or otherwise moves restlessly while the body is near the nest. This may function to shape a cup in the nest.

Sitting: The bird rests on its belly on the ground, the feet out of sight. Sitting can be done "quietly", "alertly" or while "looking around". (See "Standing")

Standing: The bird remains still, supported by both legs outstretched beneath it, and only the feet touch the substrate. The body may be parallel to the substrate, or the tail may be lower than the head. Standing may be done "quietly" (with little or no movement), "alertly" (the bird stands with eyes open and head up, and appears to be looking at or listening to something), or while "looking around".

Stepping: The Nicobar lifts its leg and puts it back down again in the same or almost the same position. Steps may occur in strings, sometimes with the legs alternating. Stepping usually occurs on a nest site or in a nest.

Threatening: The Nicobar, usually at its nest, orients towards an intruder, stretches its neck, and jabs the open beak towards the intruder. The feathers may be ruffled or the wings spread. Milder threats lacked some of these elements. (cf. Whitman 1919:142)

Treading: The Nicobar lifts its feet up and down, not necessarily alternately, while usually turning in a circle. This is done on or near a potential nest site, such as a wire platform. The Nicobar's tail is partly spread and its legs are wide apart. Treading may alternate with Head Throwing or Bowing. Males do most of the Treading. (cf. Goodwin 1970:29-30)

Tremble-Shoving: A nesting bird thrusts a stick downward, usually into the nest, while vibrating the head. Nestlings occasionally Tremble-Shoved too. (cf. Lorenz 1981:299)

Walking: The Nicobar locomotes by moving first one leg forward, then the other, at a slow to moderate pace.

<u>White-Tail Fanning</u>: The Nicobar spreads its distinctive white tail into a fan shape, and rhythmically and decisively fans it from one to sixteen times, sometimes following this with half a flap. Occasionally the tail is not completely spread. The Nicobar is usually in a tree, either standing or sitting. Members of both sexes fan their tails. (cf. Goodwin 1970:29)

<u>Wing-Leg Stretching</u>: The bird stretches one wing back, but not out, at the same time stretching the leg on the same side back. (cf. Goodwin 1970: 23; Miller & Miller 1958:7)

<u>Wing Threatening</u>: The Nicobar sometimes raises its wing or wings at the approach of another pigeon. The feathers may be erect. At highest intensity, the bird lifts the wing farther from its opponent as high as it will go and attempts to hit the other with the nearer wing. Occasionally this threat is used by a pigeon on a nest as well. (cf. Goodwin 1970: 26-28; Miller & Miller 1958:6)

<u>Yawning</u>: The Nicobar tilts its head back and opens its beak very wide. Sometimes the eyes are shut.

Appendix II

Biographies of Pigeons in the Lincoln Park Zoological Gardens Colony

The following biographies include basic demographic data on all post-fledgling Nicobar pigeons ever in the Lincoln Park Zoo colony. Pigeons are listed in order of their LCB numbers, that is, in order of their arrival in the colony. The index cross-references the pigeons by their names. Weights were measured in grams at the zoo; to avoid having to deal with long strings of decimals, I have left the weights in grams. Most of the preceding dissertation does not discuss any events after 12 August 1982. This appendix, however, is complete through 3 January 1984.

Most abbreviations in the appendix are self evident. Under the column "egg," "f" refers to a fertile egg, "i" refers to an infertile egg, and "d" to an egg destroyed before it hatched. The "death/other" column lists the date that the chick died, or was sent on loan to another zoo, or the date the egg was deserted, destroyed, or discovered to be infertile. A cross next to a date means that this breeding was not listed among records on the Nicobars, but was inferred by Kathleen Brown and me from evidence in keeper's reports, egg logs, and other Lincoln Park Zoo sources. Dates are written as month/day/year. "Band" refers to the location and color of metal and plastic bands. Prior to 1983, only plastic bands were colored. Inbreeding coefficients were calculated with the assumption that the founding members were unrelated.

Index

(cont'd.)

Name of Pigeon	LCB #'s of Pigeons With this Name
28	80-28
30	78-30, 82-307
31	78-31
35	78-35
41	77-41
42	77-42
43	75-20
45	75-20
46	81-46
47	80-47
48	78-48
53	81-153
55	78-55, 83-101
59	80-59
61	82-161
64	80-64
65	80-65
66	80-66
68	77-168
75	0-75, 82-175
76	0-76, 82-176
77	77-169
80	80-16
81	81-41

(cont'd.)

Name of Pigeon	LCB #'s of Pigeons With this Name
86	78-86
89	82-89
90	78-90
91	82-91
130	76-134
133	76-133
134	76-134

0-74

Name: K24

Band: silver/-

Sex: female

Date hatched: pre-1976

Mother:

Father:

Where hatched: wild-hatched, probably arrived 1/9/73

F = 0

Breeding History:

year	mate	egg	date laid	chick LCB	hatch date	death/other
1976	76	f		76-59/K31	5/18/76	10/17/80
1977+	76	?,d				1/77
1977	K22	f		77-41/41	5/23/77	
1977	K22	f		77-162/16	8/8/77	loan, 10/80
1978+	K22	i,d				1/9/78
1978	K22	f		78-7	2/15/78	loan, 10/78
1978	K22	f		78-31	6/6/78	loan, 10/78
1978	K29	f		78-90/90	9/9/78	loan, 9/79
1979	K22	i				2/23/79
1979	K22	?,d				5/16/79
1980	30	i				5/21/80
1980	30	?,d				5/29/80
1980	30	?,d				6/27/80
1981	48	f,d				3/30/81

(cont'd.)

year	mate	egg	date laid	chick LCB	hatch date	death/other
1981	48	?,d				5/5/81
1982	K22	f	3/11/82	82-90	4/9/82	4/11/82
1982	K22	f	4/26/82		5/25/82	5/25/82
1982	K22	i	6/23/82			8/2/82
1982	K22	f		82-259/2	9/13/82	
1983	1	i,d	1/1/83			1/7/83
1983	1	f	2/23/83			3/25/83
1983	1	f		83-25/25	5/6/83	
1983	1	f	9/2/83	83-119/19	9/29/83	

Weight: 550 g (8/25/81)

Known molts: 7/78, 11/78, 6/79, 6/80

K24 was ill and out of the aviary 8/25/81 to 9/10/81. She had been nestbuilding with K22, but was replaced by 77 the first day.

K24 was moved to the duck yard with her chick K31 on 6/17/76, and was returned to the aviary 6/20/76.

K22 and K24 deserted 78-7, who was hand-raised.

K22 and K24 deserted 78-31, who was hand-raised.

Disposition: still living at Lincoln Park Zoo

0-75

Name: 75

Band: silver/green

Sex: female

Date hatched: pre-1976

Mother:

Father:

Where hatched: possibly hatched at Lincoln Park Zoo prior to the time records were kept

F = 0

Breeding history:

year	mate	egg	date laid	chick LCB	hatchdate	death/other
1976	K22*	f			6/4/76	6/20/76

* K23 could also have been the mother of this egg.

year	mate	egg	date laid	chick LCB	hatchdate	death/other
1976	76	?,d				7/8/76
1977+	K22	I				1/13/77
1977	K22	f		77-11	2/22/77	3/10/79
1977	K29	f		77-168/68	9/15/77	
1978	76	f,d				2/4/78
1978	K22	I				3/22/78
1978	K22	f		77-48/48	6/26/78	5/10/82
1980	76	f		80-28/28	4/25/80	loan, 12/81
1980	76	f	6/13/80	80-65/65	7/11/80	loan, 12/81

(cont'd.)

Known molts: 8/78, 6/79, 8/80

75 was in the Bird House hospital or the North American cage with K29 from at least 12/1/78 to 1/11/79.

75 and K22 raised 76 and K74's chick 78-10 in 3/78.

75 had a chick in Philadelphia with 76 8/23/81. They also had an infertile egg in summer, 1981.

Disposition: loan, Philadelphia, 11/6/80

0-76

Name: 76

Band: green/silver

Sex: male

Date hatched: before 10/74

Mother:

Father:

Where hatched: possibly hatched at Lincoln Park Zoo prior to the time records were kept.

F = 0

Breeding history:

year	mate	egg	date laid	chick LCB	hatch date	death/other
1976	K24	f		76-59/K31	5/18/76	10/17/80
1976	75	?,d				7/8/76
1977+	K24	?,d				1/77
1977	K31	f		77-42/42	5/30/77	loan, 9/79
1977	K74	f		77/169/77	9/17/77	
1978	75	f,d				2/4/78
1978	K74	f		78-10/10	3/26/78	loan, 9/79
1978	K74	f		78-30/30	5/10/78	loan, 11/80
1978	K74	f		78-86/86	8/22/78	6/22/82
1979	K74	i				2/24/79
1979	K74	f		79-64	6/26/79	loan, 9/79
1980	75	f		80-28/28	4/25/80	loan, 12/81
1980	75	f	6/13/80	80-65/65	7/11/80	loan, 12/81

(cont'd.)

Known molts: 8/78, 10/78, 6/79, 6/80

76 was attacked by a cattle egret 2/2/78 while defending his nest. Afterwards, he was bleeding from his mouth, half of his cere was torn away, and his eye was swollen. Two days later the egrets destroyed the nest.

76 and K74 deserted 78-10, who was raised by K22 and 75.

76 and 75 had an infertile egg in summer, 1981, and had a chick in Philadelphia 8/23/81.

Disposition: loan, Philadelphia, 11/6/80

Death date: 11/14/81 in Philadelphia. 76 died from a head injury and a defective gizzard.

73-2

Name: K22

Band: silver/-

Sex: male

Date hatched: pre-1973

Mother:

Father:

Where hatched: wild-hatched, came to Lincoln Park Zoo 1/9/73

F = 0

Breeding history:

year	mate	egg	date laid	chick LCB	hatchdate	death/other
1976	K23 or 75	f			6/4/76	6/20/76
1976	K23	f		76-74/K74	7/31/76	1/12/80
1977+	75	i				1/13/77
1977	75	f		77-11	2/22/77	3/11/79
1977	K24	f		77-41/41	5/23/77	
1977	K24	f		77-162/16	8/8/77	loan, 10/80
1978+	K24	i,d				1/9/78
1978	K24	f		78-7	2/15/78	loan, 10/78
1978	75	i				3/23/78
1978	K24	f		78-31	6/6/78	loan, 10/78
1978	75	f		78-48/48	6/26/78	5/10/82
1979	K24	i,d				2/23/79

(cont'd.)

1979	K24	?,d				5/16/79
1979	K31	f,d				6/17/79
1980	K31	f		80-25/5	4/25/80	10/17/82
1980	K31	f	6/16/80?	80-66/66	7/12/80	6/24/82
1980	77	f			10/29/80	10/29/80
1981	77	f		81-41/81	4/1/81	loan, 12/81
1981	77	f,d			9/30/81	9/30/81
1982	K24	f	3/11/82	82-90	4/9/82	4/11/82
1982	K24	f	4/26/82		5/25/82	5/25/82
1982	K24	i	6/23/82			8/2/82
1982	K24	f		82-259/2	9/13/82	
1983	35	f	12/27/82?			1/23/83
1983	35	f	2/17/83	83-14/14	3/16/83	
1983	35	f	5/15/83		6/11/83	6/11/83
1983	35	f	8/31/83			9/8/83
1983	35	i				10/25/83

Weight: 685 g (10/28/80), 740.0 g (11/20/81)

Known molts: 6/78, 6/79, 8/80, 3/82

K22 and 75 raised 76 and K74's chick 78-10 in 3/78.

K22 and K24 deserted 78-7, who was hand-raised.

K22 and K24 deserted 78-31, who was hand-raised.

Disposition: still living at Lincoln Park Zoo

73-3

Name: K23

Band:

Sex: female

Date hatched: pre-1973

Mother:

Father:

Where hatched: wild-hatched, arrived at Lincoln Park Zoo 1/9/73

F = 0

Breeding history:

year	mate	egg	date laid	chick LCB	hatchdate	death/other
1976	K22*	f			6/4/76	6/20/76

* 75 could also have been the mother of this egg.

year	mate	egg	date laid	chick LCB	hatchdate	death/other
1976	K22	f		76-74/K74	7/31/76	1/12/80

Deathdate: 9/8/76. K23 was found dead in the free-flight area; autopsy found no cause of death.

75-20

Name: 43 or 45

Band: pink/-

Sex: male

Date hatched:

Mother:

Father:

Where hatched: came to Lincoln Park Zoo 5/14/75, probably wild-caught

F = 0

Breeding history:

year	mate	egg	date laid	chick LCB	hatchdate	death/other
1977	130	f			6/27/77	6/27/77
1978	41	f		78-55/55	7/7/78	loan, 10/80

Weight: 535 g (2/15/79)

Known molt: 9/78

On 7/28/78, a Zoo visitor attempted to steal 43.

43 was kept in Monkey House basement 12/8/78 to 2/15/79.

43 had a chick in Minnesota with 10 in late 1980 or early 1981; the chick died in a fall.

Disposition: loan, Minnesota, 9/20/79

Deathdate: 10/81 in Minnesota

76-59

Name: K31

Band: -/silver

Sex: female

Date hatched: 5/18/76

Mother: K24

Father: 76

Where hatched: Lincoln Park Zoo

F = 0

Breeding history:

year	mate	egg	date laid	chick LCB	hatchdate	death/other
1977	76	f		77-42/42	5/30/77	loan, 9/79
1978	K29	f		78-35/35	5/12/78	
1979	K22	f,d				6/17/79
1980	K22	f		80-25/5	4/25/80	10/17/82
1980	K22	f	6/16/80?	80-66/66	7/12/80	6/24/82
1980	30	i				10/17/80

Known molt: 8/78

K31 was moved to the duck yard 6/17/76 and was returned to the aviary 6/30/76.

Deathdate: 10/17/80. Autopsy found no cause of death.

76-74

Name: K74

Band: -/silver (band painted brown)

Sex: female

Date hatched: 7/31/76

Mother: K23

Father: K22

Where hatched: Lincoln Park Zoo

F = 0

Breeding history:

year	mate	egg	date laid	chick LCB	hatchdate	death/other
1977	76	f		77-169/77	9/17/77	
1978	K29	?,d				1/26/78
1978	76	f		78-10/10	3/26/78	loan, 9/79
1978	76	f		78-30/30	5/10/78	loan, 10/80
1978	76	f		78-86/86	8/22/78	6/22/82
1979	76	i				2/24/79
1979	76	f		79-64	6/26/79	loan, 9/79

Weight: 392 g at fledging

Known molt: 10/78

K74 and 76 deserted 78-10, who was raised by K22 and 75.

Deathdate: 1/12/80. K74 was found drowned in large pond. She had a superficial bruise over her left radius/ulna.

76-133

Name: K29

Band: silver/-

Sex: male

Date hatched: pre-1976

Mother:

Father:

Where hatched: presumably wild-hatched, received from Basically Birds 9/9/76

F = 0

Breeding history:

year	mate	egg	date laid	chick LCB	hatch date	death/other
1977+	130	?,d				1/77
1977+	130	?,d				2/19/77
1977	75	f		77-168/68	9/15/77	
1978	K74	?,d				1/26/78
1978	K31	f		78-35/35	5/12/78	
1978	K24	f		78-90/90	9/9/78	loan, 9/79
1980	77	f		80-64/64	6/25/80	

Weight: 595 g (10/28/80)

Known molts: 7/78, 11/78, 6/79, 8/80

K29 was in the Bird House hospital or the North American cage from at least 12/1/78 to 1/11/79.

Disposition: loan, Philadelphia, 11/6/80

76-134

Name: 130, 134

Band: blue/- or -/blue

Sex: female

Date hatched: pre-1976

Mother:

Father:

Where hatched: presumably wild-hatched, received from Basically Birds 9/9/76

F = 0

Breeding history:

year	mate	egg	date laid	chick LCB	hatchdate	death/other
1977+	K29	?,d				1/77
1977+	K29	?,d				2/19/77
1977	43	f			6/27/77	6/27/77

Deathdate: 2/23/78. 130 had aspergillosis and two tumor-like masses in the lower abdomen. She was removed to the Bird House hospital 1/31/78 and was given chloromycetin.

77-11

Name:

Band: silver/silver (painted brown)

Sex: male

Date hatched: 2/22/77

Mother: 75

Father: K22

Where hatched: Lincoln Park Zoo

F = 0

Weight: 16 g (2/22/77), 106 g (2/29/77), 530 g (3/30/77)

Known molts: 4/78, 8/78

Deathdate: 3/10/79. He had trauma to the right side of the head and secondary infection.

77-41

Name: 41

Band: yellow/silver

Sex: female

Date hatched: 5/23/77

Mother: K24

Father: K22

Where hatched: Lincoln Park Zoo

F = 0

Breeding history:

year	mate	egg	date laid	chick LCB	hatchdate	death/other
1978	43	f		78-55/55	7/7/78	loan, 10/80
1980	55	f		80-47/47	6/17/80	loan, 10/80
1980	55	?,d				10/30/80
1981	68	f		81-46/46	5/11/81	12/7/82
1982	68	?	4/29/82			5/18/82
1982	68	f	6/26/82	82-176/76	7/25/82	
1983	68	f	12/25/82?	83-3/3	1/21/83	
1983	68	?,d	4/6/83			4/24/83
1983	68	f	5/15/83	83-42/42	6/8/83	
1983	68	f		83-136/13	10/6/83	

Weight: 75 g (5/25/77), 381 g (6/23/77)

Known molts: 4/78, 8/78, 2/79, 8/80

Disposition: still living at Lincoln Park Zoo

77-42

Name: 42

Band: yellow/silver

Sex: female

Date hatched: 5/30/77

Mother: K31

Father: 76

Where hatched: Lincoln Park Zoo

F = 0.250

Weight: 15 g (5/31/77), 355 g (6/28/77)

Known molts: 8/78, 6/79

42 bred with 90 in Minnesota in 8/80.

Disposition: loan, Minnesota, 9/20/79

77-162

Name: 16

Band: silver/yellow

Sex:

Date hatched: 8/8/77

Mother: K24

Father: K22

Where hatched: Lincoln Park Zoo

F = 0

Weight: 593 g (4/4/78), 700 g (10/30/80)

Known molts: 4/78, 8/78, 5/79, 6/80

Between 3/79 and 6/79, 16 built a nest with K29. No copulation was seen, no egg was laid, although K29 stood on 16's back. In 3/80, 16 was nestbuilding (?) again with K29. In 6/80, K29 deserted 16; 16 mounted and attempted to copulate with 48. They spent two days building a nest. In 10/80, 16 and mother K24 started a nest together.

Disposition: loan, San Diego, 11/6/80

77-168

Name: 68

Band: yellow/silver

Sex: male

Date hatched: 9/15/77

Mother: 75

Father: K29

Where hatched: Lincoln Park Zoo

F = 0

Breeding history:

year	mate	egg	date laid	chick LCB	hatchdate	death/other
1979	77	f,d		twins		6/23/79
1980	77	f		80-16/80	4/22/80	1/14/81
1980	77	f		80-31	4/22/80	5/22/80
1981	41	f		81-46/46	5/11/81	12/7/82
1982	41	?	4/29/82			5/18/82
1982	41	f	6/26/82	82-176/76	7/25/82	
1982	5	f	10/16/82	82-307/30	11/13/82	
1983	41	f	12/25/82?	83-3/3	1/21/83	
1983	41	?,d	4/6/83			4/24/83
1983	41	f	5/14/83	83-42/42	6/8/83	
1983	41	f		83-136/13	10/6/83	

Weight: 530 g (10/14/77), 846.9 g (11/20/81)

(cont'd.)

Known molts: 6/78, 8/78, 12/78, 2/79, 6/79, 6/80, 4/81, 7/81, 10/81, 6/82.

After 5 died, 82-307 was pulled and substituted for 77 and 46's infertile egg, and was raised by them.

Disposition: still living at Lincoln Park Zoo

77-169

Name: 77

Band: yellow/silver

Sex: female

Date hatched: 9/17/77

Mother: K74

Father: 76

Where hatched: Lincoln Park Zoo

F = 0

Breeding history:

year	mate	egg	date laid	chick LCB	hatchdate	death/other
1979	68	f,d		twins		6/23/79
1980	68	f		80-16/80	4/22/80	1/14/81
1980	68	f		80-31	4/22/80	5/22/80
1980	K29	f		80-64/64	6/25/80	
1980	K22	f,d				10/29/80
1981	K22	f		81-41/81	4/1/81	loan, 12/81
1981	K22	f,d			9/30/81	9/30/81
1982	48	f	3/17/82	82-91/91	4/13/82	
1982	none	f	6/23/82			7/7/82
1982	46	i	10/17/82			

Weight: 507 g (10/20/77), 706.6 g (11/20/81)

Known molts: 4/78, 6/79, 8/80, 7/82

48 died in 1982 while 91 was still in the nest, and 77 finished

(cont'd.)

raising 91 herself. In June 1982, 77 laid a fertile egg in a bread basket in the screwpine. Earlier, she had been seen courting with 86, and copulating with 22. The egg was not pulled until she deserted it, and it died.

77 and 46 were given 68 and 5's fertile egg (82-307) to incubate and raise. It hatched 11/13/82. 77 continued raising 30 after 46 died 12/7/82.

Disposition: still living at Lincoln Park Zoo

78-7

Name:

Band: -/silver

Sex:

Date hatched: 2/15/78

Mother: K24

Father: K22

Where hatched: Lincoln Park Zoo

F = 0

Weight: 21 g (2/15/78), 604 g (10/4/78)

78-7 was pulled 2/8/78 because K24 deserted the nest, and was hand-raised.

Disposition: loan, Boston, 10/4/78

78-10

Name: 10

Band: yellow/silver

Sex: female

Date hatched: 3/26/78

Mother: K74

Father: 76

Where hatched: Lincoln Park Zoo

F = 0

Weight: 388 g (4/19/78)

Known molts: 2/79, 6/79

10 was deserted by original parents 3/24/78, and egg was given to surrogates 75 and K22.

10 had a chick in Minnesota with 43 in late 1980 or early 1981, but the chick died in a fall.

Disposition: loan, Minnesota, 9/20/79

78-30

Name: 30

Band: silver/yellow

Sex: male

Date hatched: 5/10/78

Mother: K74

Father: 76

Where hatched: Lincoln Park Zoo

F = 0

Breeding history:

year	mate	egg	date laid	chick LCB	hatchdate	death/other
1979	35	i				6/10/79
1980	K24	i				5/21/80
1980	K24	?,d				5/29/80
1980	K24	?,d				6/27/80
1980	K31	i				10/17/80

Weight: 685 g (10/30/80)

Known molts: 1/79, 6/79, 7/80, 9/80

Disposition: loan, San Diego, 11/6/80

78-31

Name: 31

Band: silver/yellow

Sex:

Date hatched: 6/6/78

Mother: K24

Father: K22

Where hatched: Lincoln Park Zoo

F = 0

Weight: 22 g (6/6/78), 505 g (8/4/78)

31 was deserted 5/26/78 and was hand-raised.

Disposition: loan, Boston, 10/4/78

Deathdate: 8/24/80 in Boston

78-35

Name: 35

Band: silver/yellow

Sex: female

Date hatched: 5/12/78

Mother: K31

Father: K29

Where hatched: Lincoln Park Zoo

Breeding history:

year	mate	egg	date laid	chick LCB	hatchdate	death/other
1979	30	i				6/10/79
1980	48	?,d				4/16/80
1980	48	f,d				5/9/80
1980	86	f		80-59/59	6/30/80	8/8/82
1980	86	f		80-117	11/12/80	11/14/80
1981	86	f	4/12/81			5/10/81
1981	86	?				9/26/81
1983	K22	f	12/27/82?			1/23/83
1983	K22	f	2/17/83	83-14/14	3/16/83	
1983	K22	f	5/15/83		6/11/83	6/11/83
1983	K22	f	8/30/83			9/8/83
1983	K22	i				10/25/83

Weight: 545 g (6/17/78), 738.8 g (11/20/81)

Known molts: 8/78, 2/79, 8/80, 11/81

Disposition: still living at Lincoln Park Zoo

78-48

Name: 48

Band: silver/yellow

Sex: male

Date hatched: 6/26/78

Mother: 75

Father: K22

Where hatched: Lincoln Park Zoo

F = 0

Breeding history:

year	mate	egg	date laid	chick LCB	hatchdate	death/other
1980	35	?,d				4/16/80
1980	35	f,d				5/9/80
1981	K24	f,d				3/30/81
1981	K24	?,d				5/5/81
1982	77	f	3/17/82	82-91/91	4/13/82	

Weight: 395 g (7/25/78), 645 g (10/30/80), 810.9 g (11/20/81)

Known molts: 1/79, 6/79, 3/80, 6/80, 7/81, 9/81

Deathdate: 5/10/82. 48 was seen vomiting shortly before death; autopsy showed inflammation of the esophagus. Trichomonas was possible cause of death.

78-55

Name: 55

Band: silver/green

Sex: male

Date hatched: 7/7/78

Mother: 41

Father: 43

Where hatched: Lincoln Park Zoo

F = 0

Breeding history:

year	mate	egg	date laid	chick LCB	hatchdate	death/other
1980	41	f		80-47/47	6/17/80	loan, 10/80
1980	41	?,d				10/30/80

Weight: 415 g (8/8/78), 625 g (10/30/80)

Known molts: 1/79, 4/79, 8/80

Disposition: loan, San Diego, 11/6/80

78-86

Name: 86

Band: silver/yellow

Sex: male

Date hatched: 8/22/78

Mother: K74

Father: 76

Where hatched: Lincoln Park Zoo

F = 0

Breeding history:

year	mate	egg	date laid	chick LCB	hatchdate	death/other
1980	35	f		80-59/59	6/30/80	8/8/82
1980	35	f		80-117	11/12/80	11/14/80
1981	35		4/12/81			5/10/81
1981	35	?				9/26/81
1982	59	f	5/3/82			6/9/82

Weight: 503 g (9/22/78), 840.1 g (11/20/81)

Known molts: 2/79, 8/80

Deathdate: 6/22/82. 86 was nestbuilding with 66 and 35. Two days after his death, 66 died. 86 was never autopsied. Trichomonas was possible cause of death.

78-90

Name: 90

Band: silver/yellow

Sex: male

Date hatched: 9/9/78

Mother: K74

Father: K29

Where hatched: Lincoln Park Zoo

F = 0

Weight: 515 g (10/5/78)

Known molt: 6/79

90 bred with 42 in Minnesota in 8/80.

Disposition: loan, Minnesota, 9/20/79

79-25

Name: 25

Band: yellow/silver

Sex: male

Date hatched: 3/20/75

Mother:

Father:

Where hatched: Stone Memorial Zoo, Massachusetts. 25 came from Baltimore Zoo 3/7/79.

F = 0

Weight: 543 g (3/28/79), 635 g (8/30/79)

Disposition: loan, Seattle, 9/20/79

79-64

Name:

Band:

Sex: male?

Date hatched: 6/26/79

Mother: K74

Father: 76

Where hatched: Lincoln Park Zoo

F = 0

Weight: 585 g (8/1/79)

Known molt: 3/81

Disposition: loan, Minneapolis, 9/20/79

80-16

Name: 80

Band: silver/yellow

Sex:

Date hatched: 4/22/80

Mother: 77

Father: 68

Where hatched: Lincoln Park Zoo

F = 0

Deathdate: 1/14/81. 80 had a clot in the abdominal cavity, and a large autolized white mass was attached to the spleen.

80-25

Name: 5

Band: yellow/silver

Sex: female

Date hatched: 4/25/80

Mother: K31

Father: K22

Where hatched: Lincoln Park Zoo

F = 0

Breeding history:

year	mate	egg	date laid	chick LCB	hatchdate	death/other
1982	68	f	10/16/82	82-307/30	11/13/82	

Weight: 330 g (5/21/80), 686.7 g (11/23/81)

Known molt: 4/82

After 5 died, her egg was pulled and substituted for 77 and 46's infertile egg.

Deathdate: 10/17/82. 5 was found hanging in a tree, had been seen vomiting the previous day. Trichomonas was possible cause of death.

80-28

Name: 28

Band: yellow/silver

Sex: male

Date hatched: 4/25/80

Mother: 75

Father: 76

Where hatched: Lincoln Park Zoo

F = 0

Weight: 490 g (6/4/80), 768.9 g (11/19/81)

Known molt: 11/80

Disposition: loan, San Antonio, 12/2/81

80-32

Name: 22

Band: silver/blue

Sex: male

Date hatched: 1976

Mother:

Father:

Where hatched: came to Lincoln Park Zoo 5/7/80 from Fresno Zoo

F = 0

Breeding history:

year	mate	egg	date laid	chick LCB	hatchdate	death/other
1980	23	f		80-58/1	6/27/80	
1980	23	i				10/24/80
1981	23	f		81-47/4	5/10/81	loan, 12/81
1981	23	f	8/24/81	81-153/53	9/20/81	
1982	23	f		82-89/89	4/8/82	
1982	23	?	5/19/82			5/31/82
1982	59	f	6/18/82?	82-161/61	7/18/82	
1982	23	f,d	10/11/82?			10/25/82
1983	64	f	12/25/82	83-4/4	1/23/83	5/13/83

Weight: 615 g (5/7/80), 630.3 g (12/20/81)

Known molts: 8/80, 11/81

22 was treated for bumblefoot for three days beginning 11/20/81.

Deathdate: 2/17/83. 22 was found vomiting, and died an hour later. Trichomonas and hemmorrage of the crop were found in autopsy. An outside autopsy determined that shock was the cause of death.

80-33

Name: 23

Band: silver/blue

Sex: female

Date hatched: 1976

Mother:

Father:

Where hatched: came to Lincoln Park Zoo 5/7/80 from Fresno Zoo

F = 0

Breeding history:

year	mate	egg	date laid	chick LCB	hatchdate	death/other
1980	22	f		80-58/1	6/27/80	
1980	22	i				10/24/80
1981	22	f		81-47/4	5/10/81	loan, 12/81
1981	22	f	8/24/81	81-153/53	9/20/81	
1982	22	f		82-89/89	4/8/82	
1982	22	?	5/19/82			5/31/82
1982	22	f,d	10/11/82?			10/25/82

Weight: 600 g (5/7/80), 578.6 g (11/20/81)

Disposition: still living at Lincoln Park Zoo

80-34

Name: 24

Band: blue/silver

Sex: female

Date hatched: probably 1979

Mother: probably 23

Father: probably 22

Where hatched: probably Fresno Zoo, came to Lincoln Park Zoo from Fresno Zoo 5/7/80

F = 0

Weight: 600 g (5/7/80)

Known molt: 6/80

Disposition: loan, San Diego, 11/6/80

Deathdate: 12/25/80 in San Diego. The autopsy showed visceral gout.

80-35

Name: 25

Band: blue/silver

Sex: male?

Date hatched: 7/21/79

Mother: 23

Father: 22

Where hatched: Fresno Zoo, came to Lincoln Park Zoo 5/7/80 from Fresno Zoo

F = 0

Weight: 470 g (5/7/80), 610 g (1/19/81), 465 g (6/25/82)

25 was in the Bird House hospital 6/25/82 to 8/2/82 with a broken left wing.

Although laparoscoped as a male, 25 always plays the female role in courtship.

Disposition: still living at Lincoln Park Zoo

80-47

Name: 47

Band: silver/yellow

Sex:

Date hatched: 6/17/80

Mother: 41

Father: 55

Where hatched: Lincoln Park Zoo

F = 0.250

Weight: 450 g (7/20/80)

Disposition: loan, Philadelphia, 11/6/80

80-58

Name: 1

Band: silver/yellow

Sex: male

Date hatched: 6/27/80

Mother: 23

Father: 22

Where hatched: Lincoln Park Zoo

F = 0

Breeding history:

year	mate	egg	date laid	chick LCB	hatchdate	death/other
1982	64	f,d	5/17/82			6/8/82
1982	64	f	6/27/82	82-175/75	7/24/82	
1983	K24	i,d	1/1/83			1/7/83
1983	K24	f	2/23/83			3/25/83
1983	K24	f		83-25/25	5/6/83	
1983	K24	f	9/2/83	83-119/19	9/29/83	

Weight: 430 g (7/25/80), 565 g (10/30/80), 610.5 g (11/23/81)

Disposition: still living at Lincoln Park Zoo

80-59

Name: 59

Band: silver/yellow

Sex: female

Date hatched: 6/30/80

Mother: 35

Father: 86

Where hatched: Lincoln Park Zoo

F = 0.0625

Breeding history:

year	mate	egg	date laid	chick LCB	hatchdate	death/other
1982	86	f	5/3/82			6/9/82
1982	22	f	6/18/82?	81-161/61	7/18/82	

Weight: 403 g (8/2/80), 505 g (8/14/80), 560 g (10/30/80), 610 g (11/23/81)

Known molt: 3/81

Deathdate: 8/8/82. 59 was seen vomiting frequently and looked dehydrated two days before her death. Trichomonas was possible cause of death.

80-64

Name: 64

Band: silver/yellow

Sex: female

Date hatched: 6/25/80

Mother: 77

Father: K29

Where hatched: Lincoln Park Zoo

F = 0

Breeding history:

year	mate	egg	date laid	chick LCB	hatchdate	death/other
1982	1	f,d	5/17/82			6/8/82
1982	1	f	6/27/82	82-175/75	7/24/82	
1983	22	f	12/25/82	83-4/4	1/23/83	5/13/83
1983	53	?	9/9/83			10/3/83

Weight: 330 g (7/19/80), 691 g (11/19/81)

Known molt: 3/81

In November, 1981, 64 was laparoscoped as a male.

64 continued raising 83-4 on her own after 22 died 2/17/83.

Disposition: still living at Lincoln Park Zoo

80-65

Name: 65

Band: yellow/silver

Sex: male

Date hatched: 7/11/80

Mother: 75

Father: 76

Where hatched: Lincoln Park Zoo

F = 0

Weight: 680 g (8/27/80), 824.8 g (11/23/81)

Disposition: loan, San Antonio, 12/2/81

80-66

Name: 66

Band: –/silver

Sex: female

Date hatched: 7/12/80

Mother: K31

Father: K22

Where hatched: Lincoln Park Zoo

F = 0

Weight: 415 g (8/11/80), 701 g (11/23/81)

Known molt: 4/82

Deathdate: 6/24/82. 66's mate, 86, died two days earlier. Two months after her death, 66 was autopsied. Her crop was inflamed. Trichomonas was a possible cause of death.

81-41

Name: 81

Band: silver/yellow

Sex: female

Date hatched: 4/2/81

Mother: 77

Father: K22

Where hatched: Lincoln Park Zoo

F = 0.125

Weight: 119.7 g (4/10/80), 411.3 g (5/1/81), 630.9 g (11/23/81)

At Louisiana Purchase Zoo, 81 bred with their male. She laid a fertile egg 2/18/82, which was found cracked 3/13/82, and laid an infertile egg 3/25/82. She laid another infertile egg 9/15/82.

Disposition: loan, Monroe, Louisiana, 12/2/81

81-46

Name: 46

Band: yellow/silver

Sex: male

Date hatched: 5/11/81

Mother: 41

Father: 68

Where hatched: Lincoln Park Zoo

F = 0

Breeding history:

year	mate	egg	date laid	chick LCB	hatchdate	death/other
1982	77	I				10/17/82

Weight: 412 g (6/11/81), 659.2 g (11/23/81)

77 and 46 were given 82-307 to incubate and raise in 1982.

Deathdate: 12/7/82. Earlier in the day, 46 had intermittent paralysis and seizures; vet suspected vitamin B and calcium deficiencies. Autopsy showed head injury.

81-47

Name: 4

Band: silver/yellow

Sex: female

Date hatched: 5/11/81

Mother: 23

Father: 22

Where hatched: Lincoln Park Zoo

F = 0

Weight: 463 g (6/27/81), 577.8 g (11/23/81)

Disposition: loan, New Orleans, 12/2/81

81-153

Name: 53

Band: green/silver

Sex: male

Date hatched: 9/20/81

Mother: 23

Father: 22

Where hatched: Lincoln Park Zoo

F = 0

Breeding history:

year	mate	egg	date laid	chick LCB	hatchdate	death/other
1983	64	?	9/9/83			10/3/83

Weight: 478 g (10/24/81), 507.6 g (11/19/81)

Known molt: 5/83

Disposition: still living at Lincoln Park Zoo

82-89

Name: 89

Band: silver/yellow

Sex:

Date hatched: 4/8/82

Mother: 23

Father: 22

Where hatched: Lincoln Park Zoo

F = 0

Weight: 478.6 g (5/19/82)

Disposition: still living at Lincoln Park Zoo

82-91

Name: 91

Band: yellow/silver

Sex:

Date hatched: 4/13/82

Mother: 77

Father: 48

Where hatched: Lincoln Park Zoo

F = 0.0625

Weight: 114.4 g (4/21/82), 317.4 g (5/7/82), 366.7 g (5/13/82), 552.8 g (6/7/82), 463 g (6/13/82), 598.4 g (3/12/83)

Known molt: 1/83

91 was in the Bird House hospital 6/7/82 to 6/13/82 with Trichomonas, and was treated with Flagyl.

Disposition: still living at the Lincoln Park Zoo

82-161

Name: 61

Band: silver/yellow

Sex:

Date hatched: 7/18/82

Mother: 59

Father: 22

Where hatched: Lincoln Park Zoo

F = 0

Weight: 450.1 g (8/31/82)

Disposition: still living at Lincoln Park Zoo

82-175

Name: 75

Band: yellow/silver

Sex:

Date hatched: 7/24/82

Mother: 64

Father: 1

Where hatched: Lincoln Park Zoo

F = 0

Weight: 389.1 g (8/20/82)

Disposition: still living at Lincoln Park Zoo

82-176

Name: 76

Band: silver/yellow

Sex:

Date hatched: 7/25/82

Mother: 41

Father: 68

Where hatched: Lincoln Park Zoo

F = 0

Weight: 395.6 g (8/20/82)

76 came to the ground for the first time at the age of 32 days.

Disposition: still living at Lincoln Park Zoo

82-259

Name: 2

Band: yellow/silver

Sex:

Date hatched: 9/13/82

Mother: K24

Father: K22

Where hatched: Lincoln Park Zoo

F = 0

Weight: 586 g (10/27/82)

82-307

Name: 30

Band: gold/green (gold band says 82)

Sex:

Date hatched: 11/13/82

Mother: 5

Father: 68

Where hatched: Lincoln Park Zoo

F = 0

Weight: 356.5 g (12/15/82)

5 died the day after laying 30's egg. 30 was incubated and raised first by 77 and 46, then by 77 alone after 46 died.

83-3

Name: 3

Band: pink/yellow (pink says 25)

Sex:

Date hatched: 1/21/83

Mother: 41

Father: 68

Where hatched: Lincoln Park Zoo

F = 0

Weight: 590 g (3/9/83)

83-4

Name: 4

Band: green/pink (pink band says 8)

Sex:

Date hatched: 1/23/83

Mother: 64

Father: 22

Where hatched: Lincoln Park Zoo

F = 0

Weight: 400.9 g (3/9/83)

After 22 died on 2/17/83, 4 continued raising squab on her own.

Deathdate: 5/13/83

83-14

Name: 25

Band: yellow plastic/pink metal (metal band says 14)

Sex:

Date hatched: 3/16/83

Mother: 35

Father: K22

Where hatched: Lincoln Park Zoo

F = 0

83-25

Name: 25

Band: green plastic/gold metal

Sex:

Date hatched: 5/6/83

Mother: K24

Father: 1

Where hatched: Lincoln Park Zoo

F = 0

Weight: 24.7 g (5/7/83), 78.5 g (5/11/83), 164.7 g (5/16/83), 241.1 g (5/20/83).

83-42

Name: 42

Band: blue plastic/pink metal

Sex:

Date hatched: 6/8/83

Mother: 41

Father: 68

Where hatched: Lincoln Park Zoo

F = 0

Weight: 565 g (7/27/83)

Disposition: still living at Lincoln Park Zoo

83-101

Name: 55

Band: yellow plastic 55/metal 43

Sex: female

Date hatched: pre-1965

Mother:

Father:

Where hatched: unknown. Toledo Zoo purchased her in 7/65, and she arrived at Lincoln Park Zoo 9/9/83.

F = 0

Weight: 515.7 g (9/83)

Disposition: still living at Lincoln Park Zoo

83-119

Name: 19

Band: yellow plastic/gold metal

Sex:

Date hatched: 9/29/83

Mother: K24

Father: 1

Where hatched: Lincoln Park Zoo

F = 0

Weight: 535.2 g (10/30/83)

Disposition: still living at Lincoln Park Zoo

83-136

Name: 13

Band: pink plastic/silver metal

Sex:

Date hatched: 10/6/83

Mother: 41

Father: 68

Where hatched: Lincoln Park Zoo

F = 0

Weight: 317.5 g (10/29/83), 475.8 g (11/10/83)

Disposition: still living at Lincoln Park Zoo

CURRICULUM VITA

Shauna S. Roberts

Birthdate: 17 September 1956

Birthplace: Dayton, Ohio

EDUCATIONAL EXPERIENCE:

	Major	From/To	Degree
Northwestern University Evanston, Illinois	Anthropology	9/78-6/79	M.A.
University of Pennsylvania Philadelphia, Pennsylvania	Anthropology	9/74-12/77	B.A.

PRESENTED PAPERS:

"Are Breeding Success and Mate Choice Independent in Captive Nicobar Pigeons, Caloenas nicobarica?" with Kevin J. Bell and Kathleen Brown. Presented at the Midwest Regional Animal Behavior Meeting, St. Louis, Missouri. 8-10 April 1983.

"Functions of Nest Defense in a Captive Colony of Nicobar Pigeons, Caloenas nicobarica." Presented at the Midwest Regional Animal Behavior Meeting, Champaign, Illinois. 12-14 February 1982.

PUBLICATIONS:

Sade, D. S., Chepko-Sade, B. D., Schneider, J., and Roberts, S., and others. In preparation. Basic Demographic Observations on Free-ranging Rhesus Monkeys.

HONORS AND SCHOLARSHIPS:

5/82 - 7/82	Dr. Scholl Research Fellowship. Lincoln Park Zoological Gardens, Chicago, Illinois.
9/80 - 6/81	Graduate Research Assistant Tuition Scholarship (GRATS). Northwestern University, Evanston, Illinois.
9/79 - 6/80	Departmental Assistantship. Northwestern University, Evanston, Illinois.
9/78 - 6/79	Graduate School Fellowship. Northwestern University, Evanston, Illinois.

Made in the USA
Monee, IL
04 November 2026

67382379R00125